Systems Reliability and Failure Prevention

For a listing of recent titles in the *Artech House Technology Management and Professional Development Library*, turn to the back of this book.

Systems Reliability and Failure Prevention

Herbert Hecht

Artech House
Boston • London
www.artechhouse.com

Library of Congress Cataloging-in-Publication Data
A catalog record for this book is available from the U.S. Library of Congress.

British Library Cataloguing in Publication Data
Hecht, Herbert
  Systems reliability and failure prevention.—(Artech House technology management library)
  1. Reliability (Engineering) 2. System failures (Engineering) – Prevention
  I. Title
  620'.00452

ISBN 1-58053-372-8

Cover design by Yekaterina Ratner

© 2004 ARTECH HOUSE, INC.
685 Canton Street
Norwood, MA 02062

All rights reserved. Printed and bound in the United States of America. No part of this book may be reproduced or utilized in any form or by any means, electronic or mechanical, including photocopying, recording, or by any information storage and retrieval system, without permission in writing from the publisher.
  All terms mentioned in this book that are known to be trademarks or service marks have been appropriately capitalized. Artech House cannot attest to the accuracy of this information. Use of a term in this book should not be regarded as affecting the validity of any trademark or service mark.

International Standard Book Number: 1-58053-372-8
A Library of Congress Catalog Card number is available from the Library of Congress.

10 9 8 7 6 5 4 3 2 1

*This book is dedicated to my wife, Esther,
whose encouragement made it possible.*

Contents

1	**Introduction**	**1**
2	**Essentials of Reliability Engineering**	**5**
2.1	The Exponential Distribution	5
2.2	Parameter Estimation	8
2.3	Reliability Block Diagrams	9
2.4	State Transition Methods	12
2.5	The Devil Is In the Details	16
2.5.1	Numerator of the Failure Rate Expression	16
2.5.2	Denominator of the Failure Rate Expression	16
2.5.3	Repair Rate Formulations	18
2.6	Chapter Summary	19
	References	19
3	**Organizational Causes of Failures**	**21**
3.1	Failures Are Not Inevitable	21
3.2	Thoroughly Documented Failures	22
3.2.1	Mars Spacecraft Failures	23
3.2.2	Space Shuttle Columbia Accident	26

3.2.3	Chernobyl	27
3.2.4	Aviation Accidents	29
3.2.5	Telecommunications	31
3.3	Common Threads	32
	References	34

4 Analytical Approaches to Failure Prevention — 37

4.1	Failure Modes and Effects Analysis	38
4.1.1	Overview of FMEA Worksheets	38
4.1.2	Organization of an FMEA Report	41
4.1.3	Alternative FMEA Approaches	46
4.1.4	FMEA as a Plan for Action	49
4.2	Sneak Circuit Analysis	52
4.2.1	Basics of SCA	52
4.2.2	Current SCA Techniques	55
4.3	Fault Tree Analysis	56
4.3.1	Basics of FTA	57
4.3.2	Example of FTA	58
4.4	Chapter Summary	61
	References	61

5 Testing to Prevent Failures — 63

5.1	Reliability Demonstration	64
5.2	Design Margins	66
5.3	Reliability Relevance of Tests During Development	70
5.4	Reliability Relevance of Postdevelopment Tests	76
5.5	In-Service Testing	82
5.6	Chapter Summary	85
	References	85

6 Redundancy Techniques — 87

6.1	Introduction to Redundancy at the Component Level	87

6.2	Dual Redundancy	91
6.2.1	Static and Dynamic Redundancy	91
6.2.2	Identical Versus Diverse Alternates	95
6.2.3	Active Versus Dormant Alternates	97
6.3	Triple Redundancy	98
6.3.1	TMR	99
6.3.2	Pair-and-Spare Redundancy	101
6.4	Higher-Order Redundant Configurations	102
6.5	Other Forms of Redundancy	105
6.5.1	Temporal Redundancy	105
6.5.2	Analytical Redundancy	105
6.6	Chapter Summary	106
	References	107

7 Software Reliability — 109

7.1	The Nature and Statistical Measures of Software Failures	109
7.2	Software Testing	114
7.3	Failure Prevention Practices	120
7.3.1	Requirements	120
7.3.2	Test	122
7.3.3	UML-Based Software Development	126
7.4	Software Fault Tolerance	129
7.5	Software Reliability Models	132
7.6	Chapter Summary	132
	References	133

8 Failure Prevention in the Life Cycle — 137

8.1	Life-Cycle Format and Terminology	138
8.2	Reliability Issues in Life-Cycle Phases	141
8.3	The Reliability Program Plan	148
8.4	Reviews and Audits	152

8.5	Monitoring of Critical Items	157
8.5.1	Monitoring Purchased Items	158
8.5.2	In-House Monitoring for Reliability Attainment	159
8.6	Chapter Summary	163
	References	164

9	**Cost of Failure and Failure Prevention**	**167**
9.1	Optimum Reliability	167
9.2	Time Considerations of Expenditures	170
9.3	Estimation of Cost Elements	174
9.4	A Generic Cost of Reliability Model	177
9.5	Chapter Summary	181
	References	182

10	**Cost Trade-offs**	**183**
10.1	Reliability Improvement to Meet QoS Requirements	183
10.1.1	Analysis of the Commercial Power Supply	184
10.1.2	Server Equipment Availability	187
10.2	Increasing Maintenance Effectiveness	192
10.3	Replacement of Communication Satellites	195
10.4	Chapter Summary	197
	References	198

11	**Applications**	**199**
11.1	Power Supply for Ground Communications	200
11.1.1	Framework for Power Supply Selection	200
11.1.2	Power Supply Alternatives	201
11.1.3	Evaluation of Alternatives	208
11.2	Reliability of Aircraft Electronics Bay	210
11.2.1	Primary Power Supply	210
11.2.2	Safety Critical Loads	211

11.2.3	Partial Improvement of a Function	213
11.3	Spacecraft Attitude Determination	215
11.3.1	Orthogonal Gyro Configurations	215
11.3.2	Nonorthogonal Gyro Orientation	217
11.4	Chapter Summary	219
	Reference	219

About the Author   **221**

Index   **223**

1
Introduction

The primary aim of system reliability is the prevention of failures that affect the operational capability of a system. The probability of such failures can be reduced by the following:

- Conservative design—such as ample margins, use of parts and materials with established operating experience, and observing environmental restrictions;
- Use of analysis tools and techniques—particularly failure modes and effects analysis, fault tree analysis and—for electrical components—sneak circuit analysis, followed by correcting the problem areas detected by these;
- Extensive testing—to verify design margins, toleration of environmental extremes, and absence of fatigue and other life-limiting effects;
- Redundancy—to protect against random failures by providing alternative means of accomplishing a required function.

All of these techniques, including their capabilities and limitations, are discussed in this book. In addition, there is a chapter on organizational causes of failure, a subject frequently overlooked in the reliability literature. Failures are attributed to organizational causes when a recognized cause of failure exists and known preventive measures were not installed or used.

This book was written for engineering and management professionals who need a concise, yet comprehensive, introduction to the techniques and practice of system reliability. It uses equations where the clarity of that notation is required, but the mathematics is kept simple and generally does not require calculus. Also, the physical or statistical reasoning for the mathematical model is supplied. Approximations are used where these are customary in typical system reliability practice. References point to more detailed or more advanced treatment of important topics.

Cost considerations pervade all aspects of system design, including reliability practices. Two chapters are specifically devoted to cost models and cost trade-off techniques but we also deal with economic aspects throughout the book. In addition, we recognize that reliability is just one of the desirable attributes of a product or service and that affordability, ease of use, or superior performance can be of equal or greater importance in making a project a commercial success. Thus, recommendations for reliability improvements are frequently represented for a range of achievable values.

Chapter 2 is titled "Essentials of Reliability Engineering" and deals with concepts, terminology, and equations that are familiar to most professionals in the field. It is still recommended that even the experienced reader pay a brief visit to this chapter to become familiar with the notations used in later chapters. Section 2.5, titled "The Devil Is In the Details," may warrant more than a casual glance from all who want to understand how failure rates are generated and how they should be applied. That section also discusses precautions that must be observed when comparing failure rates between projects.

Chapter 3, "Organizational Causes of Failures" is a unique contribution of this book. These causes need to be understood at all levels of the organization, but responsibility for their prevention lies primarily in the province of enterprise and project managers. Some of the failures described are due to the coincidence of multiple adverse circumstances, each one of which could have been handled satisfactorily if it occurred in isolation. Thus, the chapter illustrates the limitations of the frequently stated assumption of "one failure at a time."

Chapters 4–6 deal, respectively, with analysis, test, and redundancy techniques. We treat failure mode and effects analysis (FMEA) in considerable detail because it can serve as the pivot around which all other system reliability activities are organized. We also provide an overview of combined hardware and software FMEA, a topic that will probably receive much more attention in the future because failure detection and recovery from hardware failures is increasingly being implemented in software (and, hence, depends on failure-free software). An important contribution of Chapter 5 is design margin testing, a technique that requires much less test time than conventional reliability demonstration but is applicable only to known failure modes. Redundancy is effective primarily against random failures and is the most readily demonstrated way of dealing with these. But it is very costly, particularly where weight and power consumption must be limited, and therefore we discuss a number of alternatives.

Chapter 7 is devoted to software reliability. We discuss inherent differences in causes, effects, and recovery techniques between hardware and software failures. But we also recognize the needs of the project manager who requires compatible statistical measures for hardware and software failures. Viewed from a distance, software failures may, in some cases, be regarded as random events, and for circumstances where this is applicable, we describe means for software redundancy and other fault-tolerance measures.

In Chapter 8, we learn how to apply the previously discussed techniques to failure prevention during the life cycle. Typical life-cycle formats and their use in reliability management are discussed. A highlight of this chapter is the generation and use of a reliability program plan. We also emphasize the establishment and use of a failure reporting system.

Chapter 9 is the first of two chapters specifically devoted to the cost aspects of system reliability. In this chapter we explore the concept of an economically optimum value of reliability, initially as a theoretical concept. Then we investigate practical implications of this model and develop it into a planning tool for establishing a range of suitable reliability requirements for a new product or service. In Chapter 10 we apply economic criteria to reliability and availability improvements in existing systems.

A review of important aspects of the previous material is provided in Chapter 11 in the form of typical assignments for a lead system reliability engineer. The examples emphasize working in a system context where some components have already been specified (usually not because of their reliability attributes) and requirements must be met by manipulating a limited number of alternatives. The results of these efforts are presented in a format that allows system management to make the final selection among the configurations that meet or almost meet reliability requirements.

Reliability and cost numbers used in the volume were selected to give the reader concrete examples of analysis results and design decisions. They are not intended to represent actual values likely to be encountered in current practical systems.

The book is intended to be read in sequence. However, the following classification of chapters by reader interest may help in an initial selection (See Table 1.1). The interest groups are:

General—What is system reliability all about?

Management—What key decisions need to be made about system reliability?

Supervision—How can the effectiveness of the reliability group be improved?

Professional—How can I enhance my work and presentation skills?

Table 1.1
Guide by Reader Interest

Chapter Number and Topic	General	Management	Supervision	Professional
2 Essentials	X	O	O	X
3 Organizational causes	X	X	X	X
4–6 Techniques	O	O	X	X
7 Software reliability	O	O	X	X
8 Life cycle	X	O	X	X
9 and 10 Cost aspects	X	X	X	X
11 Applications	O	X	X	X

X: Entire chapter   O: Lead-in and chapter summary

The author drew on the experience of many years at Sperry Flight Systems (now a division of Honeywell), at The Aerospace Corporation and at his current employer, SoHaR Inc. The indulgence of the latter has made this book possible.

2

Essentials of Reliability Engineering

This chapter summarizes terminology and relationships commonly used in reliability engineering. It can be skimmed by current practitioners and gone over lightly by those who have at least occasional contact with reliability activities and documents. For all others we will try to provide the essentials of the field in as painless a manner as possible.

Stripped of legalese, the reliability of an item can be defined as (1) the ability to render its intended function, or (2) the probability that it will not fail. The aim of reliability engineering under either of these definitions is to prevent failures but only definition (2) requires a statistical interpretation of this effort such as is emphasized in this chapter.

2.1 The Exponential Distribution

In contrast to some later chapters where there is emphasis on causes of failures we are concerned here only with the number of failures, the time interval (or other index of exposure to failure) over which they occurred, and environmental factors that may have affected the outcomes. Also, we consider only two outcomes: success or failure. Initially we assume that there are no wear-out (as in light bulbs) or depletion (as in batteries) processes at work. When all of the stated assumptions are met we are left with failures that occur at random intervals but with a fixed long-term average frequency.

Similar processes are encountered in gambling (e.g., the probability of hitting a specific number in roulette or drawing a specific card out of a deck). These situations were investigated by the French mathematician Siméon Poisson (1781–1840), who formulated the classical relation for the probability of random events in a large number of trials. The form of the *Poisson distribution* used by reliability engineers is

$$P(F,t) = \frac{(\lambda t)^F}{F!} e^{-\lambda t} \qquad (2.1)$$

where λ = average failure rate;
F = number of observed failures;
t = operating time.

Reliability is the probability that no failures occur ($F = 0$), and hence the reliability for the time interval t is given by

$$R = e^{-\lambda t} \qquad (2.2)$$

This is referred to as the *exponential distribution*. An example of this distribution for $\lambda = 1$ and time in arbitrary units is shown in Figure 2.1.

At one unit on the time axis the product $\lambda t = 1$ and the reliability at that point is approximately 0.37, a value that is much too low to be acceptable in most applications. The reliability engineer must strive for a value of λ such that

Figure 2.1 Example of exponential distribution.

for the intended mission time t the product λt is much less than 1. For $\lambda t < 0.1$ (2.2) can be approximated by

$$R \approx 1 - \lambda t \tag{2.2a}$$

Thus, for $\lambda t = 0.05$, $R = 0.95$. The exact value is 0.9512. See also Insert 2.1.

The dimension of λ is 1/time. Commonly used units of λ are per hour, per 10^6 hours or per 10^9 hours. The latter measure is sometimes referred to as *fits*. A failure rate of 2 fits thus corresponds to a failure rate of 2×10^{-9} per hour.

The ordinate (reliability axis) in Figure 2.1 represents a probability measure. It can also be interpreted as the expected fraction of survivors of a population if no replacements are made for failed items. If replacements are made, the number of survivors will remain constant and so will the expected number of failures per unit time. This characteristic means that the actual time that a unit has been in service does not affect the future failure probability. For this reason the exponential distribution is sometimes called the "memoryless" distribution. When all failures are random, the failure probability is not reduced by replacement of units based on time in service. Practical systems are usually composed of at least some items that have components that are subject to wear out.

Where the failure rate depends on factors other than time it is expressed in terms of the failure-inducing factor. Thus we speak of failures per 1,000 miles for automotive systems or per million cycles for an elevator component.

For the exponential distribution the reciprocal of λ is referred to as the *mean-time-between-failures* (MTBF). A high MTBF thus corresponds to a low failure rate and is a desirable reliability attribute. The term MTBF is primarily used in applications that permit repair of a failed component. Where repair is not possible, the *mean-time-to-failure* (MTTF) is used; it is calculated in the same manner but the system level result of a failure that cannot be repaired is different. Commonly used units for the MTBF or MTTF are hours, 10^6 hours, or years.

Insert 2.1—For the Mathematically Inclined

Taking natural logs on both sides of (2.2) yields $\log_e(R) = -\lambda t$.

The series expansion of the left side is $(R-1) + \frac{1}{2}(R-1)^2 + \frac{1}{3}(R-1)^3\ldots$

As R approaches 1 the $(R-1)^2$ and higher-order terms can be neglected.

Then $R-1 = -\lambda t$ from which $R = 1-\lambda t$.

The reliability of components that involve explosive devices or other irreversible mechanisms is not a function of either operating time or of the number of cycles of use. It is simply expressed as a success probability that can be equated to the product λt in (2.1).

2.2 Parameter Estimation

When Poisson developed (2.1) for gambling situations, he was able to calculate the event probability (λt) from the structure of the game (e.g., the probability of drawing a given card from a deck of 52 is 1/52). In the reliability field we must estimate λ from experimental data (the t factor is usually specified in the statement of the problem). Because it would be difficult for each reliability engineer to collect such data, it is customary to use published summaries of part failure data, such as MIL-HDBK-217 [1] or Bellcore TR-332 [2]. These handbooks list *base failure rates,* λ_b, for each part type that have to be converted to the applicable λ for the intended use by multiplying by factors that account for the system environment (such as airborne, ground stationary, or ground mobile), the expected ambient temperature, the part's quality, and sometimes additional modifiers.

The approaches by which the published data are usually obtained carry with them significant limitations:

1. Testing of large numbers of parts—the test environment may not be representative of the usage environment.
2. Part level field failure reports—the measure of exposure (such as part operating time) is difficult to capture accurately and the failure may be attributed to the wrong part.
3. Field failure reports on systems and regression on parts population (this is explained below)—the systems usually differ in factors not associated with parts count, such as usage environment, design vintage, and completeness of record keeping.

Thus, the published failure rate data cannot be depended on for an accurate estimate of the expected reliability. It is useful for gross estimation of the achievable reliability and for assessing alternatives. Conventional reliability prediction assumes that all system failures are due to part failures, a premise that becomes fragile due to the increasing use of digital technology. Whenever the failure rate can be estimated from local data, that value will be preferable to one obtained from the handbooks.

The following is a brief explanation of the regression procedure for obtaining part reliability data from system failure reports. Assume that two systems have been in operation for a comparable period of time. Their parts populations and failure data are summarized in Table 2.1.

It is then concluded that the 10 additional digital integrated circuits (ICs) in system B accounted for the 2×10^{-6} increase in the failure rate, and each digital IC is thus assigned a failure rate of 0.2×10^{-6}. The approach can be refined by conducting similar comparisons among other systems and evaluating the consistency of the estimated failure rates for a given part.

Not all system failures are caused by parts. Other causes include unexpected interactions between components, tolerance build-up, and software faults. But because of the difficulty of associating these with numerical failure rates the practice has been to use part failure rates as a proxy, meaning that if a system has a high part failure rate it will probably have a proportionately high failure rate due to the nonanalyzed causes. A more systematic approach to accounting for nonpart-related failure is represented by the Prism program [3].

2.3 Reliability Block Diagrams

Reliability block diagrams (RBD) are used to show the relation between the reliability of a lower level element and that of a higher-level element. The lowest level element that we usually deal with is the part. We use the term *function* for the next higher element and in our examples, the element above that is the system. In many applications there will be several layers between the function and the system.

The reliability engineer considers elements to be in series whenever failure of any one lower element disables the higher level. In a simple case, the function *display lighting* consists of a switch controlling a light bulb and the RBD series relationship corresponds to series connections of the circuit, as shown in Figure 2.2(a). However, if the switch controls a relay that, in turn, controls the light bulb, the RBD shows the three elements (switch, relay, light bulb) to be in series while the schematic does not. This is shown in Figure 2.2(b), where the

Table 2.1
System Reliability Data

System	Number of Digital ICs	Number of Analog ICs	Failure Rate (10^{-6})
A	10	10	5
B	20	10	7
Difference (B – A)	10	0	2

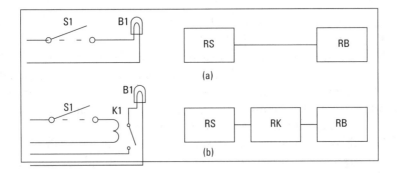

Figure 2.2 RBD for the function display lighting: (a) switch-controlled, and (b) relay-controlled.

symbols RS, RK, and RB represent the reliability of the switch, relay, and bulb, respectively.

A series relationship in the RBD implies that the higher level (function, system) will fail if any of the blocks fail, and, consequently, that all must work for the higher level to be operational. The reliability of the higher level, R, is therefore the product of the reliability of the n blocks comprising that level which is expressed as

$$R = \prod_{i=1}^{n} R_i \qquad (2.3)$$

For the configuration of Figure 2.2(a), this becomes $R = RS \times RB$, and for Figure 2.2(b) it becomes $R = RS \times RK \times RB$. In the typical case when the individual block reliabilities are close to 1, (2.3) can be approximated by

$$R \approx 1 - \sum_{i=0}^{n}(1 - R_i) \qquad (2.3a)$$

The basis for the approximation can be verified by substituting $R_i = 1-F_i$ in (2.3) and neglecting F_i^2 and higher terms. The form of (2.3a) is used in most practical reliability analysis. Thus, the failure rate (or failure probability) of a series string is obtained by summing the failure rate (or failure probability) of each of the constituent blocks.

When elements support each other such that failure of one will not disable the next higher level they are shown in parallel on the RBD. Parallel connection of parts on a circuit diagram corresponds only sometimes to a parallel RBD structure. An example of parallel connection being equivalent to parallel RBD structure is shown in Figure 2.3(a). Resistors fail predominantly by opening the circuit and only rarely by creating a short circuit. Therefore, having two resistors in parallel increases reliability, provided that either resistor can carry the current by itself and that the circuit can tolerate the change in resistance that will occur when one resistor opens.

In other cases parallel operation of parts diminishes rather than adds to reliability, as shown in Figure 2.3(b). Electrolytic capacitors fail predominantly in a short circuit mode and therefore parallel operation increases the failure probability. The reliability, R, of n parallel RBD blocks, such as the ones shown in Figure 2.3(a) is computed as

$$R = 1 - \prod_{i=1}^{n}(1 - R_i) \text{ or } F = \prod_{i=1}^{n} F_i \tag{2.4}$$

Because the decrease in reliability due to parallel operation that is illustrated in Figure 2.3(b) affects many electronic components, paralleled parts are seldom used to increase reliability. On the other hand, parallel operation (redundancy) at higher levels (assemblies or subsystems) is generally beneficial, and specific techniques are discussed in Chapter 6. Here we provide an overview of the RBD and mathematical representations of such parallel operation.

The controller subsystem shown in Figure 2.4 is redundant with two controllers, A1 and A2, being constantly powered (not explicitly shown) and receiving identical sensor inputs. The actuator is normally connected to controller A1 but if that fails it can be switched to the output of A2. The algorithm and the mechanization of switch operation are not important right now, but we recognize that the switch can fail in two ways: S'—failure to switch to A2 when it should, and S''—switching to A2 when it should not.

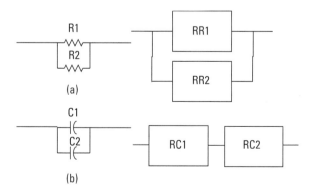

Figure 2.3 Parallel operation of parts: (a) parallel resistors, and (b) parallel capacitors.

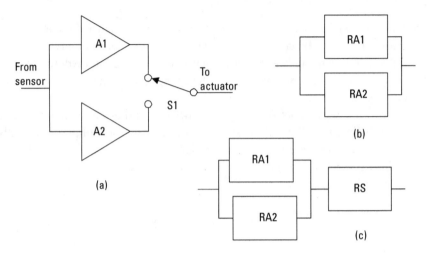

Figure 2.4 Representation of redundant controllers: (a) schematic, (b) RBD1, and (c) RBD2.

The conventional RBD does not provide a facility for distinguishing between these failure modes. Two alternatives are available: neglecting the switch failure probability, as shown in Figure 2.4(b), or postulating that any switch failure will disable the circuit, as shown in Figure 2.4(c). The former (RBD1) may be used as an approximation, where the joint failure probability of A1 and A2 is much greater than the switch failure probability. The latter (RBD2) may be used as an upper limit of the failure probability.

The difference between these alternatives will be evaluated with an example in which the failure probabilities of A1 and A2 are assumed to be identical and are designated $F_1 = F_2 = 0.001$, and the failure probability of the switch is designated $F_s = 0.0001$. Then the optimistic assumptions of RBD1 lead to a function failure probability of $F_1 \times F_2 = 10^{-6}$, while the upper limit on failure probability is $F_s + F_1 \times F_2 = 101 \times 10^{-6}$. The difference in RBD assumptions leads to a large difference in the expected reliability that will be intolerable in many applications. Therefore, better analysis procedures for this situation are discussed in Section 2.4.

2.4 State Transition Methods

These limitations of the RBD representation for some system configurations can be overcome by state tables and state transition diagrams. The failed state analysis shown in Table 2.2 provides improved discrimination between the failure modes of S1. The possible states of the controllers are listed across the top of the table (A′ denotes a failed controller), and the possible states of the switch are listed in the left column. S′ is the state in which the switch fails to select A2 after

Table 2.2
Failed State Analysis

	A1A2	A1'A2	A1A2'	A1'A2'
S				X
S'		X		X
S''			X	X

failure of A1, and S'' is the state in which the switch transfers to A2 although A1 was operational. Combinations of controller and switch states that cause failure of the function are indicated by an X.

The table permits writing the equation for failure of the controller function, F, as

$$F = Pr(S') \times Pr(A1'A2) + Pr(S'') \times Pr(A1A2') + Pr(A1'A2') \quad (2.5)$$

where Pr denotes the probability of a state. Since the probability of the non-primed states (A1 and A2) is very close to 1, (2.5) can be approximated by

$$F \approx F_{S'} \times F_1 + F_{S''} \times F_2 + F_1 \times F_2 \quad (2.5a)$$

where $F_{S'}$ and $F_{S''}$ represent the failure probabilities associated with the corresponding switch states.

We will now revisit the numerical example of Figure 2.4 with the added assumption that $F_{S'} = F_{S''} = 0.00005$. Then

$$F \sim 10^{-3} \times (0.00005 + 0.00005 + 0.001) = 1.1 \times 10^{-6}$$

In this instance, the more detailed analysis yields a result that is close to the optimistic assumption (RBD1) in Figure 2.4 but such an outcome is not assured in general.

The state transition diagram, examples of which are shown in Figure 2.5, is a very versatile technique for representing configurations that go beyond the capabilities of an RBD, particularly the evaluation of repairable systems. A non-repairable simplex function is shown in Figure 2.5(a). It has an active state (state 0) and failed state (state 1). The failure rate, λ, denotes the transition probability from state 0 to state 1. Normal operation, shown by the reentrant circle on top of state 0, is the only alternative to transition to failure and

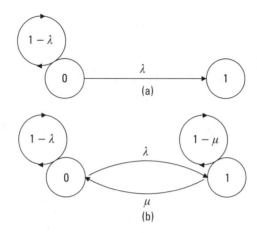

Figure 2.5 State transition diagrams: (a) nonrepairable function, and (b) repairable function.

therefore continues at the rate of $1 - \lambda$. State 1 is an absorbing state (one from which there is no exit), since this function has been identified as nonrepairable. When the transition probabilities, such as λ, are constant, the state transition diagram is also referred to as a Markov diagram after Russian mathematician Andrei A. Markov (1856–1922), who pioneered the study of chain processes (repeated state transitions).

For a repairable function, shown in Figure 2.5(b), there is a backward transition and state 1 is no longer absorbing. After having failed, a function will remain in state 1 until it is repaired (returned to state 0). The repair rate, usually denoted μ, is the reciprocal of the mean-time-to-repair (MTTR). Thus, for an element with MTTR of 0.5 hour, $\mu = 2\ hr^{-1}$.

State transition models can be used to compute the probability of being in any one of the identified states (and the number of these can be quite large) after a specified number of transitions (discrete parameter model) or after a specified time interval (continuous parameter model). The mathematics for the continuous parameter model is described in specialized texts that go beyond the interest of the average reader [4, 5]. Computer programs are available that accept inputs of the pertinent parameters in a form that is familiar to system and reliability engineers [6–8].

The discrete parameter model (in which state changes occur only at clock ticks) can be solved by linear equations, as shown here for the function of Figure 2.5(b). The probability of being in state i will be designated by Pr_i ($i = 0$, 1). Then

$$Pr_0(t+1) = Pr_0(t) \times (1 - \lambda \Delta t) + Pr_1(t) \times \mu \Delta t$$
$$Pr_1(t+1) = Pr_0(t) \times \lambda \Delta t + Pr_1(t) \times (1 - \mu \Delta t)$$
(2.6)

These relations can be evaluated by means of a spreadsheet or similar procedure, as shown in Table 2.3. We use the notation $\lambda^* = \lambda \Delta t$ and $\mu^* = \mu \Delta t$ for the transition probabilities. To obtain good numerical accuracy in this procedure, the Δt factor must be chosen such that the transition probabilities λ^* and μ^* are much less than one. At the start of operation $t = 0$, $Pr_0(0) = 1$ and $Pr_1(0) = 0$.

The probability of being in the operable state, Pr_0, is referred to as the *availability* of a repairable system. Figure 2.6 represents the results of applying the procedure of Table 2.3 with parameters $\lambda^* = 0.0005$ and $\mu^* = 0.01$. The availability starts at 1 and approaches a steady-state value, A_{ss}, given by

$$A_{ss} = \frac{MTBF}{MTBF + MTTR} \tag{2.7}$$

If we equate each interval to 1 hour, the parameters used for this computation become $\lambda = 0.0005$ ($MTBF = 2{,}000$ hrs) and $\mu = 0.01$ ($MTTR =$

Table 2.3
Calculation of Discrete Transition Probabilities

Interval	$PR_0(i)$	$PR_1(i)$
1	$PR_0(0) \times (1-\lambda^*) + PR_1(0) \times \mu^*$	$PR_0(0) \times \lambda^* + PR_1(0) \times (1-\mu^*)$
2	$PR_0(1) \times (1-\lambda^*) + PR_1(1) \times \mu^*$	$PR_0(1) \times \lambda^* + PR_1(1) \times (1-\mu^*)$
3	$PR_0(2) \times (1-\lambda^*) + PR_1(2) \times \mu^*$	$PR_0(2) \times \lambda^* + PR_1(2) \times (1-\mu^*)$

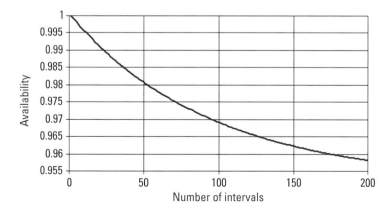

Figure 2.6 Availability computed from a discrete time model.

100 hrs). The availability calculated by means of (2.7) is 0.952, which is in agreement with the apparent asymptotic value in Figure 2.6. Availability is an important figure of merit for communication, surveillance, and similar service systems. Several examples are covered in Chapter 11. Air traffic control systems, for example, typically aim for an availability of essential services of at least 0.9999999, sometimes written as 0.9_7.

2.5 The Devil Is In the Details

The mathematical rigor of the proceeding sections must be tempered by uncertainties in the meaning of the critical parameters λ and μ. Both are customarily expressed as quotients (failures or repairs per unit time) but the content of the numerator and the denominator is far from standardized. Differences in the interpretation of these quantities can introduce errors that far exceed those caused by use of an inappropriate model. The "details" referred to in the title of this section are particularly important when failure statistics from two projects are compared, or when the time history of failures from a completed project is used to forecast the trend for a new one. We will first examine the numerator and denominator of λ and then the overall expression for μ.

2.5.1 Numerator of the Failure Rate Expression

Differences between the user's and the developer's view of what constitutes a "failure" are encountered in many situations but are particularly prominent in software-enabled systems (see Section 8.1). The user counts any service interruption as a failure, while the developer views removal of the causes of failures as a primary responsibility. Thus, multiple failures caused by a software error, an inadequately cooled microchip, or an overstressed mechanical part are single events in the eyes of the developer. No conclusions can be drawn when failure statistics arising from service interruptions are compared to removal of causes of failures. A further issue in counting failures in the numerator of the failure rate expression is the scoring of "Retest OK" (RTOK) or "Could Not Determine" (CND) failure reports. These can arise from physical intermittent conditions but also from "solid" failures in components that need to be active only under infrequently encountered conditions of use. RTOK and CND failures can account for up to one-half of all reported failures in some military equipment. If one failure reporting system includes these and another one does not, comparisons between the two will obviously be flawed.

2.5.2 Denominator of the Failure Rate Expression

The denominator of the failure rate expression should be a measure of exposure to failure. The customary measure of exposure in cars is the number of miles driven, and in printers it is the number of copies made, but in most cases it is time. And different interpretations of what aspect of time is the failure-inducing process can cause difficulties. Suppose that we have established that projects A and B count failures in the same way and express failure rates in units of 10^{-6} per hour. Is not this sufficient to make them candidates for comparison? Not necessarily. The most common interpretation of time is calendar time—720 hours per month and approximately 8,650 hours per year. If the usage of both projects is uniform, then we may have a valid comparison. But if project A serves a mountain resort that is open only 3 months of the year and project B serves a metropolitan area, failure rates that are based on calendar time would not be comparable.

When equipment with added capabilities is introduced by operating it alongside established units (serving the same purpose), the new equipment will see only occasional use because personnel is familiar with the old one. In time, the added capabilities will be recognized and use of the new equipment will increase. It is not uncommon to see an increase in the reported failure rate because higher usage causes additional failure exposure (even if it had been powered on while not in use) and because failures are now observed in some operating modes that had never been exercised. Some of these uncertainties can be removed by using run-time indicators or computer logging to capture utilization data.

A typical experience from the introduction of new equipment is shown in Figure 2.7. While the number of failures per calendar time increases, the number of failures per operational time (of the new equipment) decreases. There is no "true" reliability trend in this example. The number of failures per month is significant because it represents service interruptions. The decreasing trend in the number of failures per operating time indicates that the equipment

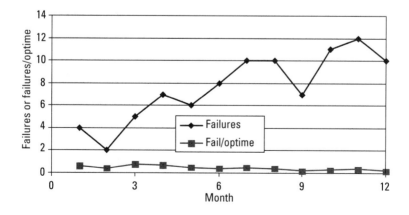

Figure 2.7 Failures in calendar time and operating time.

is getting more reliable and that a reduction in failures per month can be expected once the operating time reaches a steady level.

Commonly used units of time and their advantages and limitations are shown in Table 2.4.

There is no single "best" denominator for the failure rate function. The most important conclusion is that failure rate data sources should be used only when the time basis is known and is applicable to the environment into which the information will be imported.

2.5.3 Repair Rate Formulations

Although the repair rate, μ, is the preferred parameter for state transition analysis, it is more convenient to use its reciprocal, $1/\mu$, the repair time, in the following discussion. It is almost always expressed in hours (and these are calendar hours) but there are different ways of starting and stopping the clock, as shown in Figure 2.8.

The most easily defined time interval is the time to restore service (TTRS). As shown in Figure 2.8, it starts when (normal) operation is interrupted and it stops when operation resumes. In one classification, the TTRS is divided into administrative, location, and repair times, as indicated by the three rows of lines above the TTRS interval. Representative activities necessary to restore service are shown by vertical labels above each line segment. Only one dispatch segment is shown in the figure, but several would normally be required where parts have to be requisitioned.

TTRS is the significant quantity for availability calculations. Repair time is the most significant quantity for computing cost of repair and staffing projections. Here again it is the "details"—the use of approximately the same name for different measures of repair that must be taken into account when comparing statistics from different projects.

Table 2.4
Units of Time for Failure Reporting

Unit	Discussion
Calendar time	Universally available but suitable only for continuously running items
Adjusted calendar time	Universally available, suitable for items running by workweek, etc.
Power-on time	Suitable where power is a major stress-inducing factor
Operating time	Generally good indicator of failure exposure, requires a monitor
Execution time	Good indicator for computer-based functions, requires logging
Number of executions	Compensates for differences in execution speed of computers

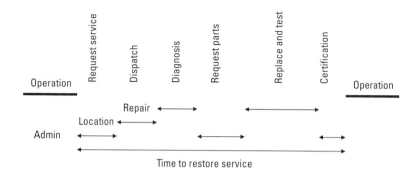

Figure 2.8 Time elements for repair.

2.6 Chapter Summary

In this chapter, we examined basic tools of reliability engineering for modeling random failure processes. The exponential distribution and the reliability block diagram representation will be encountered many times in later chapters and in practically any book or article on systems reliability. State tables and state transition diagrams are also common tools but are more specialized, and their use depends on equipment configuration and application environment.

Let us remind the reader that the exponential distribution applies only to random failures. But the mathematical simplicity of the exponential failure law motivates its use where failures are not strictly random: the early life failures or life-limited parts, mixed populations of parts where some may be failing due to deterministic causes, and even to software failures, as we shall see in Chapter 7.

In Section 2.5, "The Devil Is In the Details," we hope we alerted the reader to pitfalls that may confound the novice and have claimed victims among experienced practitioners as well.

The failure concept used in this chapter is primarily suitable for hardware components. In Chapter 3, we will deal with a broader spectrum of failures, many of which cannot be directly traced to a part failure or to any physical failure mechanism.

We also want to caution that reading and understanding this chapter has not made you a reliability engineer. But when you have a discussion with one, it may enable you to ask the right questions.

References

[1] Department of Defense, *Military Handbook, Reliability Prediction of Electronic Equipment*, MIL-HDBK-217F, December 1991. This handbook is no longer maintained by the Department of Defense but is still widely used.

[2] Bellcore *TR-332*, now available as Telcordia SR-332 *Reliability Prediction for Electronic Equipment* from Telcordia Technologies, Morristown, NJ.

[3] Denton, W., "Prism," *The Journal of the RAC*, Third Quarter 1999, Rome NY: IIT Research Institute/Reliability Analysis Center, 1999, pp. 1–6. Also available at http://rac.iitri.org/prism/prismflyer.pdf.

[4] Trivedi, K. S., *Probability and Statistics With Reliability, Queuing, and Computer Science Applications*, Englewood Cliffs, NJ: Prentice-Hall, 1982.

[5] Siewiorek, D. P., and R. S. Swarz, *The Theory and Practice of Reliable System Design*, Bedford, MA: Digital Press, 1982.

[6] Sahner, R.A., K. S. Trivedi, and A. Puliafito, *Performance and Reliability Analysis of Computer Systems: An Example-Based Approach Using the SHARPE Software Package*, Boston, MA: Kluwer Academic Publishers, 1995.

[7] Tang, D., et al., "MEADEP: A Dependability Evaluation Tool for Engineers," *IEEE Transactions on Reliability,* December 1998.

[8] *Application Note: Modeling Operational Availability Using MEADEP*, SoHaR Inc., June 2000, (available from www.Meadep.com).

3

Organizational Causes of Failures

Common sense (reinforced from time to time by advertising) tells us that the reliability of our vehicles, appliances, and services (utilities, banking) depends on the reliability efforts made by the vendor. These efforts, in turn, are largely dependent on what customers demand. Our toleration of unreliable equipment and services has radically diminished in recent years and, in response, most vendors have been able to improve reliability. In this chapter, we examine the tension between economy of resources and the (sometimes) very high cost of failures. We look at applications associated with high reliability requirements or expectations and analyze failures experienced there. We concentrate on the management and organizational aspects of the failures; physical causes of failures will be discussed in later chapters.

3.1 Failures Are Not Inevitable

We value safety and reliability, and we take pride in our accomplishments in these areas. When we see deficiencies, we demand that those responsible take corrective action. Frequently these demands are pressed by lawyers (who do not seem to mind that there are occasional lapses of responsibility at high corporate levels). Our professional schools teach good design practices, government at all levels enforces safety requirements in our buildings and vehicles, and developers of equipment for critical applications have standard practices in place to avoid known causes of failure. One way of evaluating the results of these efforts is through accident statistics: for example, the death rate from nonvehicular accidents declined from 94 per 100,000 population in 1907 to 19 in 1997.

Vehicular accidental deaths declined by 13% in the decade between 1989 and 1999, even though there were more vehicles and more miles driven [1].

At the personal level, experience has taught us that mechanical, electrical, and electronic devices in our environment are reliable and safe. Thus, we do not hesitate to drive over a bridge on our way to work or to walk under it on a Sunday afternoon; we fly to a destination on another continent and expect to arrive in time for a meeting; and we depend on our electronic address book to help us retrieve phone numbers.

But in spite of the application of good design practices, and the growing oversight by government agencies, we have had spectacular failures in Mars probes, commercial aircraft, phone systems, and nuclear power plants. The investigations presented in this chapter show that failures can be attributed to specific design deficiencies, lapses in review, or negligence in maintenance. But if we want to prevent failures, we must recognize that the aspects of human nature that were encountered in the investigation of these accidents are still with us. Also, we must become aware that frequently there are conflicts between the function and performance of the item being designed and the demands of reliability and safety. An automobile designed like a tank would be safer than most current models but it would meet neither the consumer's transportation needs nor his or her ability to pay. Thus, failures cannot be prevented in an absolute sense but must be controlled to be within limits dictated by consumer demands, government regulation or, as we will see in Chapters 9 and 10, economic considerations.

3.2 Thoroughly Documented Failures

The reason for examining admittedly unique and rare failures is that records of their occurrences, causes and, in some cases, remedies tend to be thorough, have been reviewed by experts, and are in the public domain. In the much more frequent incidents of traffic light outages, slow service from Internet service providers, or lines at the bank because "the computer is down," the cause of the failure is usually not known with such certainty and, even if it is, will not be divulged to the public. The more common type of failure is of more concern to us in our personal life and is also, in many cases, the target of our professional efforts.

Common sense suggests that the same processes that are observed in these well-documented incidents are also at work in the more common ones. Along the same lines, in the lead-in to this chapter we used statistics for *fatal* accidents to show that public demands and policy can reduce the likelihood of the ultimate failure. Both the base (population) and number of accidental deaths are known with fair accuracy. The same data for failures of gas furnaces or elevator controls is either not available at all or cannot be compared over a significant time span.

3.2.1 Mars Spacecraft Failures

In late 1999 two spacecraft of the NASA/Jet Propulsion Laboratory (JPL) Mars Exploration Program failed in the final stages of their intended trajectory. NASA headquarters appointed a Mars Program Independent Assessment Team (MPIAT). The following details of the causes of failure are excerpted from that team's report [2].

The Mars Climate Orbiter (MCO) was launched in December 1998 to map Mars' climate and analyze volatiles in the atmosphere. The spacecraft was intended to orbit Mars for approximately 4 years and act as a telemetry relay for the Polar Lander (discussed below). The spacecraft was lost in September 1999 due to a navigation error. Spacecraft operating data needed for navigation was provided by Lockheed Martin in English units rather than in the specified metric units. This is, of course, a serious error but it is not the first mismatch of units in a space program. The checking, reviews, and testing that are normally a part of the readiness procedures for a space launch should have detected the error and caused it to be corrected.

The MPIAT report states:

> In the Mars Climate Orbiter mission, the system of checks and balances failed, allowing a single error to result in mission failure. Multiple failures in system checks and balances included lack of training, software testing, communication, and adherence to anomaly reporting procedures, as well as inadequate preparation for contingencies. All of these contributed to the failure.

The Mars Polar Lander (MPL), launched in January 1999, was lost during landing on the planet in December 1999. The spacecraft also carried two microprobes that were intended to penetrate the Martian soil. These probes constituted a separate mission, Deep Space 2 or DS-2, which was also lost. There was no provision for landing-phase telemetry; this was a marginally acceptable design decision for the Lander but was judged to be a serious deficiency for future mission planning. The cause of the failure had to be established by inference rather than direct observation. The following excerpt from the MPIAT report refers to the intended landing sequence shown in Figure 3.1.

> The most probable cause of the MPL failure is premature shutdown of the lander engines due to spurious signals generated at lander leg deployment during descent. The spurious signals would be a false indication that the lander had landed, resulting in premature shutdown of the lander engines. This would result in the lander being destroyed when it crashed into the Mars surface. In the absence of flight data there is no way to know whether the lander successfully reached the terminal descent propulsion phase of the mission. If it did, extensive tests have shown that it would almost certainly have been

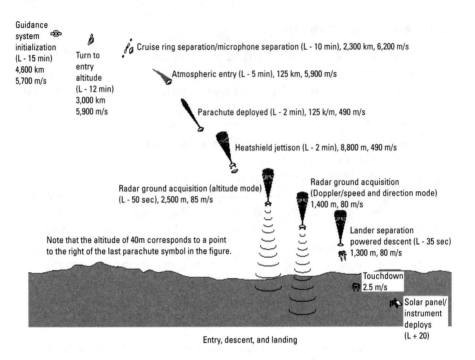

Figure 3.1 Mars Polar Lander sequence of operations. (*Courtesy:* NASA/JPL/Caltech.)

lost due to premature engine shutdown. [The figure] provides a pictorial of the MPL entry and landing sequence. Lander leg deployment is at Entry + 257 seconds. Initial sensor interrogation is at an altitude of 40m. It is at this point that spurious signals would have prematurely shut down the lander engines. As with MCO, the most probable cause of failure of the Mars Polar Lander are inadequate checks and balances that tolerated an incomplete systems test and allowed a significant software design flaw to go undetected.

The "incomplete system test" in this quote is a reference to finding a wiring error in the system test but never repeating the test after the presumed correction of the wiring error. The "significant software design flaw" is the failure to protect against spurious signals from lander legs in the engine shutdown program. Protection against spurious signals (usually called debounce check, consisting of repeated sampling of the signal source) is a common software practice when mechanical contacts are used for actuation of a critical function. The hardware design group knew that spurious signals could be generated when the legs were being extended and it could not be established whether this information had been shared with the software designers.

A mismatch of units of measurement was the immediate cause for the failure of MCO, and lack of software debounce provisions was the immediate cause of the failure of MPL. Are these separate random events? The quoted excerpts from the MPIAT report speak of inadequate checks and balances for both missions, and thus we are motivated to look for common causes that may have contributed to both failures. The MPIAT report provides an important clue by comparing the budget of the failed missions with that of the preceding successful Pathfinder mission, as shown in Table 3.1.

It is immediately apparent that NASA administration demanded (or appeared to demand) a "two for the price of one" in the combined MCO and MPL budget. That MCO and MPL were each of comparable scope to the Pathfinder can be gauged from the "Science and Instrument Development" line. But even if the combined MCO and MPL budget is compared with that of Pathfinder, there is an obvious and significant deficiency in the "Project Management" and "Mission Engineering" lines. Thus, shortcuts were demanded and taken.

The MPIAT concluded that, "NASA headquarters thought it was articulating program objectives, mission requirements, and constraints. JPL management was hearing these as nonnegotiable program mandates (e.g., as dictated launch vehicle selection, specific costs and schedules, and performance requirements)."

The MPIAT report concludes that project management at both JPL and Lockheed Martin was faced with a fixed budget, a fixed schedule, fixed science requirements, and the option of either taking risks or losing the projects altogether. Risks are being taken in most major projects; there were probably many

Table 3.1
Budget Comparison (all amounts in 1999 million $)

Budget Element	Pathfinder	Combined MCO & MPL
Project management	11	5
Mission engineering and operations development	10	6
Flight system	134	133
Science and instrument development	14	37
Rover	25	
Other	2	7
Total	196	188

instances in the Mars missions where risks were taken and did not result in a disaster. The pervasive risks accepted by the project in reducing the extent of reviews, retests, and other checking activities were not communicated to others in the organization. This lack of communication between JPL (the risk managers) and the NASA Office of Space Science (the budget managers) was identified as an important contributor to the failures in the MPIAT report.

3.2.2 Space Shuttle Columbia Accident

NASA's space shuttle Columbia lifted off from Cape Canaveral, Florida, on January 16, 2003, and it burned up due to structural damage on an orbiter leading wing edge during reentry on February 1, killing all seven astronauts on board. The wing edge was damaged by a piece of foam insulation from the external fuel tanks that broke off about 80 seconds after lift-off and caused a hole in the wing edge. During reentry fiery hot gases entered through this hole and destroyed electronics and structural members, making the shuttle uncontrollable and leading to the loss.

The investigation established that:

- Impact of foam insulation on the orbiter was an almost routine incident during launch; remedial measures were under consideration but had been postponed because the impact had not previously posed a hazard (though damage had been noted when the orbiter returned).
- The piece of foam that struck Columbia was unusually large (about twice as large as any previously observed piece) and struck a particularly sensitive part of the wing; these facts were known from movies that became available the day after the lift-off.
- Once the damage occurred, the loss of the orbiter was inevitable but a crew rescue mission could have been set in motion as late as 4 days after launch; this step was not taken because of management belief that the foam could not hurt the reinforced carbon composite material, a belief that was shown to be completely wrong by tests conducted after the accident.

The following organizational causes are believed to be involved:

1. Stifling of dissenting or questioning opinions during management meetings and particularly during flight readiness reviews, almost a "the show must go on" atmosphere.
2. Contractors received incentives for launching on time, thus causing neglect of tests or other activities that would delay a launch schedule.

3. Safety personnel were part of the shuttle organization and were expected to be "team players."

3.2.3 Chernobyl

The 1986 accident at the nuclear power station at Chernobyl in the Ukraine must be regarded as one of the most threatening events in recent history that did not involve hostile acts. The following account of the accident, probable causes, and remedial actions is excerpted from the Internet site of the World Nuclear Association [3].

On April 25, prior to a routine shutdown, the reactor crew at Chernobyl-4 began preparing for a test to determine how long turbines would spin and supply power following a loss of the main electrical power supply. Similar tests had already been carried out at Chernobyl and other plants, despite the fact that these reactors were known to be very unstable at low power settings. A diagram of the essential features of the RBMK reactor is shown in Figure 3.2. Significant differences from U.S. reactor designs are the use of graphite moderator in the Russian design (versus water) and direct access to fuel elements. Both features facilitate recovery of weapon grade nuclear material.

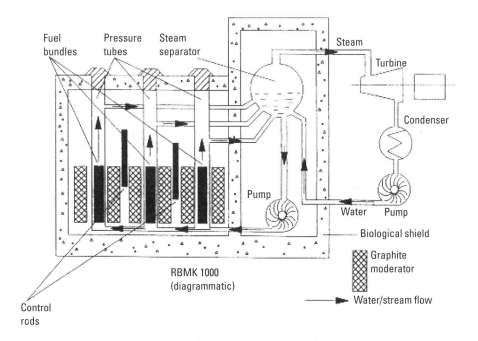

Figure 3.2 Essential features of the Chernobyl reactor. (*From:* http://www.world-nuclear.org/info/chernobyl/chornobyl.gif.)

A series of unusual operator actions, including the disabling of automatic shutdown mechanisms, preceded the attempted test early on April 26. As flow of coolant water diminished, power output increased. When the operator moved to shut down the reactor from its unstable condition arising from previous errors, a peculiarity of the design caused a dramatic power surge. The fuel elements ruptured and the resultant explosive force of steam lifted off the cover plate of the reactor, releasing fission products to the atmosphere. A second explosion threw out fragments of burning fuel and graphite from the core and allowed air to rush in, causing the graphite moderator to burst into flames.

A basic cause of the accident was that the reactor was unstable at low power settings. This instability created a point where, owing to the reactor's positive void coefficient, a further increase of steam generation would lead to a rapid increase in power. To conduct the planned test, the operators had to be extremely alert and trained in contingencies. The test had been scheduled to be conducted by the day shift but due to a power shortage the authorities in Kiev requested at the last minute that power be maintained at least at 1,600 MW, whereas the test was to be conducted at 700 MW. The test was therefore delayed into the night shift and several errors in operation of the controls occurred that might have been avoided by the day shift. A critical part of the test procedure was that the operator gives the "hold power at required level" signal. Either that signal was not given or the regulating system failed to respond to it.

The World Nuclear Association report concludes that the accident was caused by "a flawed Soviet reactor design coupled with serious mistakes made by the plant operators in the context of a system where training was minimal. It was a direct consequence of Cold War isolation and the resulting lack of any safety culture."

Another indication of the multiple causes that were involved can be found in the following safety measures taken in the former Soviet block after the Chernobyl accident:

- No further installation of RBMK reactors and the shutdown of some others;
- Education on the consequences of mistakes in reactor operations;
- Installation of improved information boards in control rooms;
- Improved operator preparation;
- Procedures to prevent the disablement of safety systems;
- Rules that made a violation of procedures a criminal offense;
- Installation of mechanical control rods to prevent them from being withdrawn too far in the RBMK reactors;
- Required provisions for more control rods;

- Automatic shutdown if an operator pulls out too many control rods;
- Changes to the composition of the fuel to increase the effectiveness of emergency shutdown.

3.2.4 Aviation Accidents

The descriptions of the following accidents were obtained from the National Transportation Safety Board (NTSB) Web site in February 2002 [4].

3.2.4.1 American Airlines Flight 1420

On June 1, 1999, at 2350:44, American Airlines flight 1420, a McDonnell Douglas DC-9-82, crashed after it overran the end of runway 4R during landing at Little Rock, Arkansas. Flight 1420 departed from Dallas/Fort Worth with 2 flight crew members, 4 flight attendants, and 139 passengers aboard and touched down in Little Rock at 2350:20. After running off the end of the runway, the airplane struck several tubes extending outward from the left edge of the instrument landing system localizer array, located 411 feet beyond the end of the runway; passed through a chain link security fence and over a rock embankment to a flood plain; and collided with the structure supporting the runway 22L approach lighting system. The captain and 10 passengers were killed; the first officer, the flight attendants, and 105 passengers received serious or minor injuries; and 24 passengers were not injured. The airplane was destroyed by impact forces and a postcrash fire.

The NTSB determined that the probable causes of this accident were the following:

- The flight crew's failure to discontinue the approach when severe thunderstorms and their associated hazards to flight operations had moved into the airport area.
- The crew's failure to ensure that the spoilers had extended after touchdown.

Contributing to the accident were the flight crew's:

1. Impaired performance resulting from fatigue and the situational stress associated with the intent to land under the circumstances;
2. Continuation of the approach to a landing when the company's maximum crosswind component was exceeded;
3. Use of reverse thrust greater than 1.3 engine pressure ratio after landing.

3.2.4.2 TWA Flight 800 Midair Explosion

TWA flight 800 departed from John F. Kennedy International Airport (JFK) for Paris on July 17, 1996, about 2019, with 2 pilots, 2 flight engineers, 14 flight attendants, and 212 passengers on board. About 2031 it crashed into the Atlantic Ocean. All 230 people on board were killed, and the airplane was destroyed. Visual meteorological conditions prevailed for the flight, which operated on an instrument flight rules flight plan.

The NTSB determined that the probable cause of the TWA flight 800 accident was an explosion of the center wing fuel tank (CWT), resulting from ignition of the flammable fuel/air mixture in the tank. The source of ignition energy for the explosion could not be determined with certainty, but, of the sources evaluated by the investigation, the most likely was a short circuit outside of the CWT (investigators found deteriorated insulation in a wire bundle that carried both high voltage wiring and the low-voltage fuel instrumentation) that allowed excessive voltage to enter it through electrical wiring associated with the fuel quantity indication system.

Contributing factors to the accident were the following:

1. The design and certification concept that fuel tank explosions could be prevented solely by precluding all ignition sources (a single point failure mechanism).
2. The design and certification of the Boeing 747 with heat sources, particularly the air conditioning system, located beneath the CWT with no means to reduce the heat transferred into the CWT or to render the fuel vapor in the tank nonflammable. Fuel is generally nonflammable at temperatures below 90°F. It is estimated to have been at 107°F at the time of the accident.

3.2.4.3 Hageland Aviation, Barrow, Alaska

On Saturday, November 8, 1997, the pilot, who was also the station manager, arrived at the airport early to prepare for a scheduled commuter flight to a nearby village, transporting seven passengers and cargo. There was heavy frost on vehicles the morning of the flight, and the lineman servicing the aircraft reported a thin layer of ice on the upper surface of the left wing. The pilot was not observed deicing the airplane prior to flight and was described by other employees as in a hurry to depart on time. The pilot directed the lineman to place fuel in the left wing only, which resulted in a weight imbalance of between 450 and 991 lbs. The first turn after takeoff was into the heavy left wing. The airplane was observed climbing past the end of the runway and descending vertically into the water. All eight persons aboard were killed. Examination of the aircraft revealed no mechanical causes for the accident.

The NTSB determined the following as probable causes:

- Pilot's disregard of lateral fuel loading limits;
- Improper removal of frost prior to takeoff;
- Failure to anticipate possible stall/spin.

Contributing causes were the following:

1. Self-induced pressure by the pilot to take off on time;
2. Inadequate surveillance of operations by company management.

3.2.5 Telecommunications

Failure investigations of telephone and other communication services are usually of much more limited scope than the ones covered in the previous examples. We cite two reports here that are in the public domain and are probably representative of failures in that field.

3.2.5.1 Fire at the Medford, Oregon, Telephone Switching Center

A fire in the Medford switching facility disrupted the (commercial) electric power supply. But do not worry; we have diesel generators just for that purpose. When the fire department arrived at the scene, they immediately demanded that the diesel engines be shut down. But do not worry; we have batteries. Word of the fire at the phone exchange and the arrival of multiple fire engines caused a significant increase in telephone traffic and the batteries were soon depleted. The Amateur Radio Emergency Service (ARES) was activated and handled communications that permitted a center in Portland to supply needed blood to a hospital in Klamath Falls.

By design, the switching facility could handle double failures (commercial power and either generator or battery) and still remain operational. The fire represented a single event that disabled all three power sources. The consequences of a single event disabling all power sources should have been foreseen but was not.

3.2.5.2 Flood in Salt Lake City Central Office

On November 18, 2000, the Terminal Doppler Weather Radar (TDWR) used by Federal Aviation Administration (FAA) traffic control personnel failed due to a communication problem. The FAA contractor, MCI-Worldcom, investigated and reported that numerous channel (telephone connection) banks were interrupted when a contractor caused a flood at the U.S. West central office in Salt Lake City. The phone company was able to establish an alternate path by

2:05 a.m. on November 19. The long recovery time was partly due to lack of requirements for dealing with anomalous conditions. Manual switching to satellite communications would have reduced the outage time [5].

Communications for air traffic control are designed to be resilient under failures of communication lines and cables. In this case, the flood in the central office disabled multiple cables and thus disrupted the data flow to the weather radar. Better training of traffic control personnel (including the use of satellite communication) could have mitigated the effect of this failure.

3.3 Common Threads

The failures previously described occurred in diverse environments and involved different failure mechanisms. Nevertheless, we want to see whether there are common threads among them that may provide guidance to prevent failures in other instances. For this purpose, we summarized probable causes in Table 3.2.

The primary and contributing causes were established by the investigating team, except for the telecommunications examples. In general, a primary cause is one that, by itself, had high probability of causing the failure, while a contributing cause did not. That dividing line can get blurred, and in the context of this chapter we note that both primary and secondary causes are often associated with organizational deficiencies. To look at the brighter side: failures can be prevented by addressing these causes at the organizational level.

When we look at the causes of failure described in Table 3.2, the following factors are present in most of them:

- Multiple primary causes;
- Deviations from required procedures or good practice;
- The perceived pressure (on individuals or an organization) to meet budget, schedule, or another imposed objective.

Multiple primary causes are evident in all cases except for TWA 800 and the telecommunication failures. In the TWA accident, the acceptance of a single-point failure mechanism is identified as a contributing cause. In the telecommunication outages, the design was intended to provide multiple independent channels but overlooked events that could affect all of them due to a single cause. These findings are not surprising because we design critical systems such that no single creditable failure will cause a catastrophic system event. When we overlook or accept single-point failures, it is usually only a question of time until a catastrophe occurs.

The second bullet item really attests to the frailties of human beings. We may be trained, taught, and examined on the required procedures to manage a

Table 3.2
Summary of Probable Causes of Failures

Incident	Probable Primary Cause(s)	Contributing Cause(s)
Mars Climate Orbiter	Error in navigation units; lack of checks and balances	Inadequate resources; perceived pressure to stay within budget
Mars Polar Lander	Lack of debounce software; incomplete testing; lack of checks and balances	Inadequate resources; perceived pressure to stay within budget
Chernobyl	Unstable reactor design; operational errors	Lack of training and contingency planning; perceived pressure to complete the test
American Airlines 1420	Flight into poor weather; failure to extend spoilers	Crew fatigue; exceeding crosswind limit; exceeding thrust reverser limits
TWA 800	Short circuit in fuel tank wiring	Heat source close to fuel tank; tolerating single-point failure mechanism
Hageland Aviation	Lateral fuel unbalance; improper deicing; not anticipating stall conditions	Pilot under schedule pressure; inadequate surveillance of operations by company management
Telecommunications (both)	Overlooked common cause for failures assumed to be independent	Lack of contingency planning and training

project or operate a critical piece of equipment, and yet in some situations we will either forget or deliberately omit an essential step. In well-managed projects, there are checks and balances so that no single decision can lead to a catastrophic outcome. In the NASA Mars missions, the absence of checks and balances was identified as a primary cause. In the Chernobyl accident, it can be inferred from the fact that a night shift crew was charged with conducting a potentially dangerous experiment. In the American Airlines accident, the routine cross-checking of pilot and copilot actions was ineffective because both were at the limit of their endurance.

The perceived pressure to achieve organizational goals may be a root cause for the deviations from established procedures already discussed. It was specifically identified in the analysis of the Mars spacecraft failures where policy statements by NASA headquarters were taken as unalterable requirements by the JPL project office. The project office could meet the requirements only by taking shortcuts that then led to the mission failures. The Chernobyl operators and the American Airlines and Hageland pilots saw it as their duty to meet schedules. Occasions in which organizational pressures cause such dire consequences are rare, but these pressures can be seen in many more mundane failures, and they represent a significant limitation on achieving high reliability.

Reliability requires resources, and there are limits to the resources that can be made available. The decision whether to accept the risk of failure or to authorize resources for risk reduction must be made at a high level. In the organizational pressures already described, individuals thought that resources were nonnegotiable *and that therefore it was useless to inform the higher levels of the risks that were being taken.* It is unfortunately true that some management practices discourage dissemination of "bad news," but this does not relieve those who know that lack of resources causes unusual risk-taking of the obligation to inform those who control the resources, particularly if the risk is due to deviating from established procedures. In the analysis of the failures, this is frequently referred to as a "communication problem." Chapters 10 and 11 describe trade-off techniques that facilitate communication between designers and those responsible for the budget.

Nonmonetary resources also impose limits on reliability. Prominent among these are the following:

- For aircraft and other vehicular designs—weight and volume;
- For space and missile systems—weight and power;
- For commercial products—appearance and time to market;
- For communication systems—capacity and coverage.

There is a bright side to all these limitations: As we learn more about both organizational and technical causes of unreliability, we are able to reduce the probability of failure by concept and design, and this enables us to avoid the resource gobblers of redundancy and test selection (rejecting products that do not meet specifications or have inadequate margins). Thus, we are able to produce aircraft and spacecraft that can carry more payload rather than having to allocate weight for triple and quadruple redundant control and instrumentation systems. To achieve these goals we need good communication about the risk of failure at all levels of the organization and analysis of failures as they occur.

References

[1]   National Safety Council, *Injury Facts*, Washington D.C., 2000 Edition.

[2]   NASA Press Release 00-46, March 26, 2000.

[3] http://www.world-nuclear.org/info/Plant Safety, Accidents/Chernobyl Accident.

[4] http://www.ntsb.gov, in February 2002.

[5] Federal Aviation Administration, Federal Aviation Traffic Management Center, *Daily Morning Briefings Summary Database,* Washington, D.C., 2000.

4

Analytical Approaches to Failure Prevention

The analytical techniques discussed in this chapter are usually the most cost-effective means of failure prevention. They can be carried out early in the development and thus minimize rework and retesting. Analysis is cheaper than modeling and much cheaper than testing. Analytical approaches to failure prevention fall into two broad classes:

1. Analyses performed to demonstrate that the performance requirements will be met (and therefore, by implication, that the item will not fail in normal use). Examples of these analyses are stress and fatigue analysis for mechanical items, worst-case analysis and thermal analysis for electronic circuits, and stability analysis for control systems.
2. Analyses performed to demonstrate that safety and reliability requirements are met. Examples of these are failure modes and effect analysis, fault tree analysis, and sneak circuit analysis.

The former are highly domain specific, as can be seen from the examples. The techniques that are required, and to an even greater extent the procedures, vary widely even among such functionally similar items as an electromechanical relay, a solid state relay, and a digital decoder. These analyses are typically performed by the designer rather than a reliability engineer, but the latter should be aware of the results of the analyses. Some examples in Section 8.3 refer to these analyses but details on their conduct must be obtained from specialized references. Within the scope of this chapter, we focus on the second, reliability-specific, type of analyses.

4.1 Failure Modes and Effects Analysis

This is a mainstay of analytical techniques for failure prevention. For each failure mode, effects are evaluated at the local, intermediate ("next higher"), and system level. Where the system level effects are considered critical, the designer can use this information to reduce the probability of failure (stronger parts, increased cooling), prevent propagation of the failure (emergency shut-off, alarm), or compensate for the effect of the failure (standby system). We first provide an overview of FMEA worksheets, followed by the organization of an FMEA report. Then we discuss variations of the established techniques and their areas of applicability and describe current developments that facilitate the inclusion of ICs and software in a FMEA. The applications of FMEA and its extension to failure modes, effects and criticality analysis (FMECA) are described in Section 4.1.4.

4.1.1 Overview of FMEA Worksheets

A formal FMEA methodology was developed in the late 1960s [1] but informal procedures for establishing the relation between part failures and system effects date back much further [2]. Yet FMEA is not ready for the rocking chair. In 2000 the Society of Automotive Engineers (SAE) published specialized FMEA procedures for the automotive industry [3]. FMEA is widely used in the process industry in support of safety and reliability [4]. It can help:

- Component designers identify locations where added strength (or derating), redundancy, or self-test may be particularly effective or desirable;
- System engineers and project managers allocate resources to areas of highest vulnerabilities;
- Procuring and regulatory organizations determine whether reliability and safety goals are being met;
- Those responsible for the operations and maintenance (O&M) phase plan for the fielding of the system.

A formal FMEA is conducted primarily to satisfy the third bullet, an imposed requirement that does not originate in the development team and thus is sometimes given low priority. Informal studies along the lines of the first two bullets are frequently undertaken in support of the development, but they are seldom published as legacy documents. The fourth bullet is perhaps the most neglected use of the FMEA. But with growing recognition that O&M costs usually overshadow those associated with the acquisition of systems, that issue deserves emphasis. O&M activities are addressed in Section 8.2.

The following are the essential concepts of the FMEA process:

1. Parts can fail in several modes, each of which typically produces a different effect. For example, a capacitor can fail open (usually causing an increased noise level in the circuit) or short (which may eliminate the entire output of the circuit).
2. The effects of the failure depend on the level at which it is detected. Usually we distinguish between three levels that are described later in this chapter.
3. The probability and severity of in-service failures can be reduced by monitoring provisions (built-in test, supervisory systems).
4. The effects of a failure can be masked or mitigated by compensating measures (redundancy, alarms).

FMEA worksheets present the information on each of these in a standardized tabular format, and this enables reviewers to identify and ultimately correct deficiencies. The organization of worksheets into an FMEA report is discussed later. The FMEA worksheet from MIL-STD-1629A [5] is shown in Figure 4.1. Although this standard is no longer active, its provisions serve as a generic format for FMEA documentation.

The worksheet consists of a header and the tabular section. The left part of the header identifies the system and component for which this sheet has been prepared, and the right header shows when and by whom. The failure mode identification number (ID) is shown in the first column in the tabular section. It is usually a hierarchical designator of the form *ss.mm.cc.ff*, where *ss* is an integer representing the subsystem, *mm* an integer representing the major component, *cc* an integer representing the lower level component, and *ff* an integer or alphabetic suffix representing the failure mode. The ID is a convenient way of referencing failure modes and is referred to in later examples.

The next two columns contain textual descriptions either for an "item" that is a single part (switch, integrated circuit, or a gear) or a function (bias supply, gear reduction). The item descriptions may apply to more than one failure

System _____ Indenture level _____ Reference drawing _____ Mission _____				Failure Mode and Effects Analysis					Date _____ Sheet ____ of ____ Compiled by _____ Approved by _____			
Identification number	Item/functional identification (nomenclature)	Function	Failure modes and causes	Mission phase operational mode	Failure effects			Failure detection method	Compensating provisions	Severity cases	Remarks
					Local effects	Next higher level	End effects				

Figure 4.1 Example of an FMEA worksheet.

mode and ID. The entries in the failure modes and causes column are unique to the ID. Where the item is a part, the failure mode description can be very simple, like "open" or "short," and the cause may be described as random failure, overload, or environmental degradation. Where the item consists of more than a single part, the description may need to be more detailed, and the cause is usually traced to parts, such as "R1o, R4o, C2s," denoting that an open failure in resistors R1 and R4 or a shorting failure in capacitor C2 all produce this failure mode for the circuit.

Where the phase of operation causes significantly different effects for a given failure mode, the phase descriptions are entered into the next column and separate row entries are made for each mission phase. An example is a failure mode that produces only minor effects when the system is in maintenance mode but major effects when it is in operation. This column can be omitted where the effects do not vary significantly between operating modes.

Failure effects are usually described at three levels:

1. Local—the function in which the failed part is located, such as an oscillator, a gear reduction, or a counter;
2. Next higher level (NHL)—usually a partition of the system that furnishes outputs (services) recognized by the system user, such as total flow calculation, input valve positioning, or status display; frequently, the NHL is a user replaceable assembly;
3. System—the highest level of a given development project, such as artillery fuze, plant communication system, or dashboard display.

The next column deals with the detection method. Failures should be detected at the lowest possible level because that permits a very specific response and reduces the probability that other failures may interfere with the detection. Thus, a local failure effect description such as "stops oscillator" should lead to providing a means for detecting the presence of oscillator output. This is usually simpler (in mechanization) and involves less delay than detecting failures at the NHL in a function dependent on the oscillator. Detection at the local level also points directly to the failed component and thus conserves important maintenance resources.

The column labeled "Compensating Provisions" is used to record all means that may be available to avoid or reduce the severity of system effects due to this failure mode. Typical entries are switching to redundant components, use of a less accurate measurement (barometric altitude instead of radar), or avoidance of some operations (placarding).

The expected system effect and the available compensating provisions enter into the assessment of the failure severity in the next-to-last column. A representative scale for this assessment is shown in Table 4.1.

Table 4.1
Typical Severity Categories

Category	Severity	Definition
I	Catastrophic	A failure that may cause death or system loss (e.g., aircraft, launching system, mission control).
II	Critical	Failures that may cause severe injury, major property damage, or system damage that will result in loss of the intended mission.
III	Marginal	A failure that may cause (1) minor injury, minor property damage, or minor system damage that will result in delay, (2) loss of availability of asset, (3) loss of mission-operating margin without significant reduction in probability of mission success.
IV	Minor	A failure not serious enough to cause injury, property damage, or system damage.
	None	Primarily used for failures that do not cause any system effect in the designated phase of operation.

A scale of 10 severity categories, with 10 being the most severe one, is commonly used in the automotive industry. In applications where the system effects do not ever constitute a threat to life, the highest two categories can be reassigned to (I) complete system loss or extensive property damage and (II) substantial system degradation or property loss. For the treatment of failure probability in the worksheet see Insert 4.1.

The last column, titled "Remarks," can be used to record other information relevant to a particular failure mode. Typical uses are to list:

- Reference documents that explain the assignment of NHL or system failure effects that may not be obvious;
- More severe effects that may result if another failure is also present;
- Precursor events that may induce this failure mode.

4.1.2 Organization of an FMEA Report

The information contained in the worksheets needs to be organized so that it can be quickly reviewed by decision makers who may be concerned with the suitability of: (a) the design for the intended purpose; (b) the product in a specific environment; and (c) the product in the usage environment against regulatory requirements. In addition to these gatekeepers, whose approval is crucial to the success of a product, the FMEA report also serves a number of support functions, such as test and maintenance planning, spares provisioning and positioning, and preparation for user and maintainer training.

> **INSERT 4.1—Failure Probability Estimates in the FMEA Worksheet**
>
> The conventional FMEA worksheet shown in Figure 4.1 does not require failure probability information. It is intended to focus attention on failures with the highest severity, regardless of expected frequency of occurrence. The latter factor enters into criticality analysis that is usually combined with FMEA to form FMECA (Failure Modes, Effects and Criticality Analysis). In the methodology of MIL-STD-1629, expected frequency of occurrence can be expressed in qualitative terms (frequent, reasonably probable, occasional, remote, and extremely unlikely), or quantitatively from reliability predictions (see Chapter 2).
>
> The qualitative approach requires less labor but makes higher demands on the experience and integrity of the analyst. Computer-based FMEA tools reduce the labor for the quantitative approach and it is now the more common one; it is used in Tables 4.2 through 4.4. The quantitative approach is also the preferred one when FMEA is combined with maintainability analysis.

Although the preparation of the report is essentially a bottom-up activity, the report is organized top-down (starts with the system level), and this eventual top-down view must be anticipated during the preparation. System level effects (at the very least, all associated with a category I or II severity) and NHL effects that contribute to them should be defined before starting the worksheets. The NHLs must be identified, and a rationale for the assignment of effects to local levels should be established. Failure to accomplish this will result in inconsistent entries for these effects on the worksheets and require backtracking to produce a workable top document. Hazards analysis documentation is a good source for identification of significant failure effects.

Figure 4.2 shows the table of contents for a typical FMEA report; this one covers a missile. The introduction usually describes the formal basis for the analysis (contract number, requestor, dates), the participants in the analysis, and highlights of the results. The second chapter presents details of FMEA procedures that may have been either prescribed by the contract or adapted from a standard or best-practices document. It then identifies the sources for failure modes and failure rates information. In the format shown in Figure 4.2, the drawings that were used are referenced in later sections. Alternatively, and particularly in smaller projects, they may be identified in the procedures chapter.

The first actual FMEA data is presented in the System Chapter (the missile level summary in Figure 4.2) and is preceded in our format by a listing of system level drawings and a graphic representation of the interactions between the major components, as shown in Figure 4.3. The numbers along the connecting

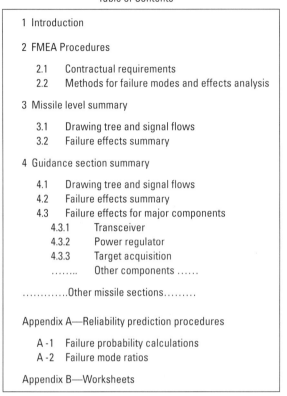

Figure 4.2 Table of contents for an FMEA report.

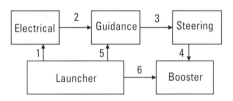

Figure 4.3 Missile level components.

lines are keyed to tables that list the individual signals in each path. The arrows indicate the dominant direction of information flow, but usually there are also some signals that are transmitted in the opposite direction. Discussion of the power distribution and the signal flow is part of this section of the report, since this is essential for assigning failure effects.

The system level summary of failure effects for the flight phase is shown in Table 4.2. An equivalent table lists failure probabilities for the prelaunch phase.

Table 4.2
Flight Phase Failure Effects Summary

Block ID	Description	Failure Probability by Severity			
		I	II	III	IV
1	Guidance section		5.123	1.408	0.039
2	Electrical section		30.627	14.949	58.431
3	Steering section		0.442	0.473	1.364
Total			36.192	16.830	59.83

The presentation of this data is accompanied by a discussion of major contributors to the failure probability and potential for reliability improvement. In this case, much of the failure probability of the electrical section was due to batteries, and selection of a more reliable battery type was recommended. Although the three blocks included in this summary do not give rise to failures in severity category I, the column for that category is retained in the summary because failures in other missile sections that are not shown here may cause category I failures.

The system failure effects summary is followed by a summary for each of the major blocks (subsystems). The format for these is identical to that of the system level section of the report: drawing tree, block diagram, and discussion of power and signal flow, followed by a summary of the failure effects. An example of this for the guidance section (flight phase) is shown in Table 4.3.

Below this level, it can be helpful to summarize failure effects by the function or service that is affected, as is shown in Table 4.4 for the receiver. This presentation permits reviewers to assess the effects in a more detailed manner than the four severity categories.

Loss of up- and downlink is a more severe failure than the loss of either link alone. As shown in the table, it carries a very small failure probability and thus is not a major concern. The table suggests that the highest priority for reliability improvement in the transceiver should be given to failures that cause loss of the uplink. The failure probability entries in Table 4.4 can be obtained directly from worksheets or, particularly for major components, from intermediate summaries generated for functional partitions.

The top-down presentation in the FMEA report permits reviewers to determine whether:

Table 4.3
Flight Phase Failure Effects for Guidance Section

Blk ID	Description	Failure Probability by Severity			
		I	II	III	IV
1.1	Receiver		2.495	0.107	
1.2	Encoder/decoder		0.364	0.020	
1.3	Power regulator		0.230	0.047	
	Other components		2.034	1.234	0.039
Total			5.123	1.408	0.039

Table 4.4
Flight Phase Failure Effects for Receiver

Next Higher Level Effect	Failure Probability by Severity			
	I	II	III	IV
Loss of up- and downlink		0.016		
Loss of uplink		2.199		
Degraded uplink			0.106	
Loss of downlink		0.280		
Launch abort			0.001	
Total		2.495	0.107	

- Safety (particularly for category I failures) and reliability objectives have been met;
- There are areas where failure probabilities need to be reduced or failure detection capability needs to be increased;
- Additional analysis has to be performed (e.g., where results do not agree with expectations).

4.1.3 Alternative FMEA Approaches

From the preceding paragraphs, it might be assumed that generating an FMEA report can be a costly activity, and that assumption cannot be dismissed. The following are ways to reduce the required effort:

- Use computer-based FMEA tools;
- Analyze at the functional rather than part level;
- Limit the scope of an FMEA.

Computer-based tools reduce analysts' workload by offering templates for entering information, furnishing libraries of failure rates and failure mode ratios for commonly used parts (and permitting the user to create additional libraries), and aggregating and summarizing worksheet data for presentation in a top-down report, as previously described [6, 7]. A sample worksheet screen from one of the commercially available software packages is shown in Figure 4.4. Each component shown with a square symbol preceding the name can be expanded by means of a mouse click (i.e., see the receiver component in the figure). The capability for hierarchical contraction and expansion permits generation of top-level reports, as shown in Tables 4.2 and 4.3, directly from the worksheets.

Another report-generation capability of this tool is illustrated by the Pareto distribution report shown in Figure 4.5. All parts included in the FMEA are listed in the order of their failure rate contribution. The last column lists the

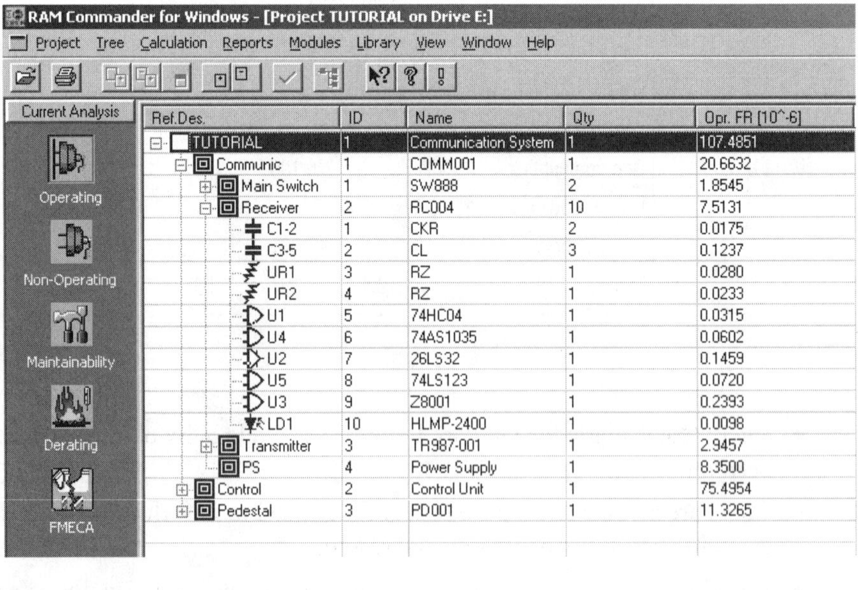
RAM Commander for Windows - [Project TUTORIAL on Drive E:]
Project Tree Calculation Reports Modules Library View Window Help
Current Analysis
Operating
Non-Operating
Maintainability
Derating
FMECA
Ref.Des. ID Name Qty Opr. FR [10^-6]
TUTORIAL 1 Communication System 1 107.4851
Communic 1 COMM001 1 20.6632
Main Switch 1 SW888 2 1.8545
Receiver 2 RC004 10 7.5131
C1-2 1 CKR 2 0.0175
C3-5 2 CL 3 0.1237
UR1 3 RZ 1 0.0280
UR2 4 RZ 1 0.0233
U1 5 74HC04 1 0.0315
U4 6 74AS1035 1 0.0602
U2 7 26LS32 1 0.1459
U5 8 74LS123 1 0.0720
U3 9 Z8001 1 0.2393
LD1 10 HLMP-2400 1 0.0098
Transmitter 3 TR987-001 1 2.9457
PS 4 Power Supply 1 8.3500
Control 2 Control Unit 1 75.4954
Pedestal 3 PD001 1 11.3265

Figure 4.4 RAM Commander screen.

RAM Commander for Windows - [Pareto Report By Name for Project TUTORIAL Drive E:]

Report Edit View Window Help

PN	Qty	Total Failure rate	Item Failure rate contribution	Cumulative contribution
KB003	1	17.000	15.816%	15.816%
CD98AB1	1	14.430	13.425%	29.241%
F99	1	12.581	11.705%	40.946%
MB00887	1	12.000	11.164%	52.111%
HDD002	1	11.484	10.685%	62.795%
Power Supply	1	8.350	7.769%	70.564%
CRT001	1	7.000	6.513%	77.076%
ANT555	1	6.658	6.195%	83.271%
MOT978	1	4.500	4.187%	87.457%
Z8001	10	2.393	2.226%	89.684%
D9S	1	2.379	2.213%	91.897%
26LS32	11	1.581	1.470%	93.367%
CL	33	1.282	1.193%	94.560%
CTRL001	1	1.000	0.930%	95.491%
RZ	32	0.817	0.760%	96.251%
74LS123	11	0.771	0.717%	96.968%
DP2	4	0.625	0.581%	97.549%
74AS1035	10	0.602	0.561%	98.110%
8086A	2	0.402	0.374%	98.484%
74HC04	10	0.315	0.293%	98.777%
D15S	1	0.241	0.224%	99.001%

Figure 4.5 Pareto distribution of part failure probability.

cumulative contribution up to this part. Thus, it can be seen that the first four parts in this listing account for more than 50% of the failure rate. Any reliability improvement effort should therefore concentrate on these parts. Other uses of the Pareto distribution are discussed in Section 5.4.

The tools furnish a large selection of output formats to meet most reporting requirements. Effective use of the tools requires training, and the greatest benefits are achieved by specialist groups that can maintain proficiency.

Another cost-reduction possibility is to perform the FMEA at the functional level rather than by parts. Most electronic components can be partitioned into functions such as power supply, timing, input filtering, signal processing, and output control. Similar functional partitions can be established for mechanical and electromechanical equipment. If failure modes for each of the functions can be defined, the analysis can be performed with much less effort than from a parts list. Also, for software and high-density ICs, the parts approach is generally not applicable and most organizations use the functional method for these. However, the disadvantages of this approach must also be considered. Table 4.5 compares significant attributes of the conventional (parts) and functional FMEA.

The most significant deficiency of the functional approach is the lack of a completeness criterion. Any reasonably complex component can be partitioned

Table 4.5
Comparison of Parts and Functional FMEA

Attribute	Parts Approach	Functional Approach
Prerequisite	Detailed drawings	Functional diagram
Completeness criterion	Based on parts list	Difficult to establish
Knowledge of failure modes	Past experience	Examination of function
Failure mode probability	Compiled sources	Must be estimated
Usual progression	Bottom-up	Top-down
Relative cost	High	Low

in more than one way; hence, what one analyst may consider a complete functional breakdown may be just a top-level breakdown to another one. The lack of functional failure mode data, similar to data that exists in the open literature for many parts, may be a serious handicap in acceptance of a functional FMEA by a regulatory body. Also, functional partitioning usually concentrates on operational functions and omits functions associated with loss of power, diagnostics, and maintenance.

The labor required for an FMEA can sometimes be considerably reduced by limiting the scope of the analysis to an area of concern that comprises the most critical failures of a device. As shown in Figure 4.6, the area of concern is the intersection of system vulnerabilities with the failure modes of the device under analysis. The area of concern must be established through discussion between the reliability engineer and the system engineer. Sometimes the system user must also be involved. The critical failures are usually not the most commonly encountered ones because the equipment users are familiar with the commonly encountered ones and know how to cope.

Very significant savings were achieved by applying this principle in an FMEA for an electronic artillery fuze. The function of the fuze is to detonate the shell at a preset time after launch when it is over the target. The fuze contains a number of mechanical and electrical safeguards against detonation prior to being fired, as well as electronic safeguards against premature detonation after firing while it might be over friendly troops.

Failure to detonate was the most prominent failure mode of the electronics, but this was just one contributor to the probability of the fuze not destroying the target. Failure to detonate was sufficiently frequent that the probability could be assessed from firing tests. Thus, the fuze's system vulnerability was

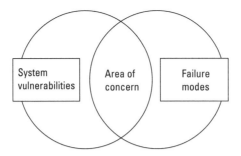

Figure 4.6 Identifying the area of concern.

identified to be premature detonation after firing, which exempted many of its functional portions. A careful FMEA of all circuit elements, the failure of which could cause premature detonation, was completed at approximately one-quarter of the cost for a complete FMEA of the electronics.

FMEA and Model-Based Development

As previously indicated, the parts approach to FMEA is generally not applicable to software and large ICs, and the functional approach is frequently unsatisfactory because there is no unique way for functional partitioning of a large IC or computer program. There have been attempts to overcome these difficulties by partitioning output properties (pins for ICs and output variables for programs) [8]. These provide clearly defined failure modes but they are not complete. They neglect internal failures in ICs that can affect more than a single output pin and programming errors that affect more than a single variable.

A more promising approach is to capture the functions as they are incorporated into the design, and this has been made practicable in model-based design, particularly when it uses techniques of the Uniform Modeling Language (UML) [9]. Functions are specified in a structured form, such as use-case diagrams [10], and the design is generated automatically (or in computer-aided fashion) from these specified functions. Details about generating a software FMEA from UML-based programs are discussed in Chapter 7.

4.1.4 FMEA as a Plan for Action

The most frequently encountered reason for performing an FMEA is that it is required as part of a project review, either in conformance to a standing organizational practice, or because of customer or regulatory demands. Project management needs to take action when the FMEA shows shortcomings, and this subsection discusses lists and summaries that have been found helpful in formulating a plan to deal with weaknesses that have been identified in the FMEA. Also, we discuss briefly the use of the FMEA as a starting point for project support activities. The use of the FMEA to initiate project improvement planning is shown in Figure 4.7.

The top row lists activities that receive essential information input from the FMEA but usually process this information in their own formats. The lower tier shows five critical item lists that are directly derived from the FMEA and may be furnished as part of the FMEA effort. Table 4.6 describes how these lists are used.

The preferred action in all cases is to improve the design so that there are no entries at all in these lists. Where this cannot be accomplished, mitigating measures should be provided (e.g., easy maintenance for high failure rate items).

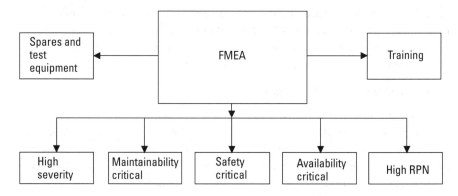

Figure 4.7 Documents generated from the FMEA.

Table 4.6
Generation and Usage of Critical Item Lists

Description	Generated From	Typical Usage	Example Thresholds
High severity	Severity category	Safety analysis	Category I & II (military), 9 & 10 (automotive)
Maintainability	Detection method	Maintainability analysis	No low-level detection
Safety critical	Severity and remarks	Safety analysis	See note*
Availability critical	Fail rate and maintainability analysis	Operational availability	Fails monthly and long replacement time
High-risk priority number (RPN)	Fail rate, severity and low detectability	Mitigation effort in automotive industry	RPN > 80 (each factor 1 – 10)

*Note: This includes failure modes that are not high severity by themselves but that remove a layer of protection (e.g., a high temperature monitor). These conditions are usually noted in the "Remarks" column of the worksheet.

It is seen that maintainability critical items are those that have a high failure rate and fail in a mode that is difficult to detect. Another consideration for the maintainability critical item list is difficulty of access. Availability critical items are a subset of the maintainability critical items that have the highest failure rates. Another factor for this list is access to specialized repair personnel and spares. It is seen that these lists deal with conditions that constitute direct impairments of the operational effectiveness of the equipment. They are therefore the subject of careful review by project decision makers.

The last designations to discuss are high criticality and high RPN. These are alternative summaries of the top issues from the FMEA that should be addressed by project management. The high criticality designation is usually preferred by military agencies, while the high RPN presentation is used in the automotive and process industries. The following are differences between the two approaches:

- Criticality is based on severity and frequency only; RPN also uses difficulty of detection;
- Criticality uses a two-dimensional graphic presentation; RPN uses algebra (multiplication).

An example of a criticality ranking graph is shown in Figure 4.8. Dark shading represents the area of highest criticality and successively lighter shadings represent lower rankings. Because this graph is intended to focus attention on the most urgent problem areas, two shadings are usually sufficient. A representation similar to that in Figure 4.8 is also used to define safety integrity levels or safety importance levels (both with the abbreviation SIL) in process controls and for the certification of some flight control equipment.

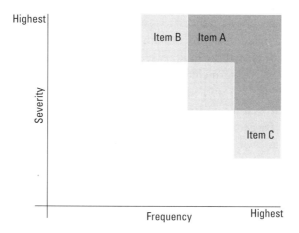

Figure 4.8 Criticality categories.

The specific items (A, B, and C) shown in the figure will motivate project management to lower their criticality, usually by moving to the left in the diagram (reducing the frequency of occurrence). The most effective means of accomplishing this is usually to add redundancy or to provide detection of precursor events (see Chapter 6).

The RPN is computed from

$$RPN = D \times F \times S \qquad (4.1)$$

where D = difficulty of detection, F = failure probability, and S = severity. All factors are expressed on a scale of 1 (lowest) to 10. The highest possible RPN score is therefore 1,000. Acceptable upper limits are application dependent and usually in the range of 50–100. Thus, a failure mode with severity 8 typically cannot tolerate a $D \times F$ product of more than 10.

4.2 Sneak Circuit Analysis

Sneak circuit analysis (SCA) is a detailed examination of switching circuitry that controls irreversible functions such as electroexplosive devices (squibs) and latches. The former Military Standard for Reliability Programs (MIL-STD-785B) defines SCA (Task 205) as a task " ... to identify latent paths which cause occurrence of unwanted functions or inhibit desired functions, assuming all components are functioning properly." This last clause is very important. Whereas FMEA considers failures in elements of a design, SCA assumes that all parts are functioning in accordance with their specification but looks for conditions that can cause undesirable system states even in the absence of failures.

The Mercury-Redstone launch failure in 1961 that incinerated the missile and launch facility was found to be due to a sneak circuit that unintentionally shut off the rocket motor immediately after ignition, causing the missile to fall back on the launch pad. A number of other mishaps due to sneak circuits in missiles and torpedoes caused the military services and NASA to require formal procedures to prevent these incidents. A *Sneak Circuit Analysis Handbook* was published in 1970 [11] and 10 years later, the requirements for SCA had become sufficiently common to lead to the Navy publication of a *Contract and Management Guide for Sneak Circuit Analysis* [12]. The following subsection introduces the basics of SCA, followed by a discussion of current SCA practices.

4.2.1 Basics of SCA

Most sneak circuits reported from production systems are too complex to describe in this introductory discussion. However, the essential characteristics of a sneak circuit can be explained with a hypothetical example of an aircraft cargo door release latch, as shown in Figure 4.9.

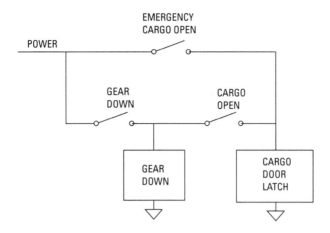

Figure 4.9 Sneak circuit in cargo door latching function.

To prevent unintended opening of the cargo door in flight, the normal cargo door control (CARGO OPEN) is powered in series with the GEAR DOWN switch. This permits routine opening on the ground. But there can be emergencies that require jettisoning cargo, and to be prepared for these there is an EMERGENCY CARGO OPEN switch that may be guarded with a safety wire to prevent its unintended operation. Now assume that an in-flight emergency exists that requires opening the cargo door. The initial action is to flip the normal CARGO OPEN switch and nothing happens (since the GEAR DOWN switch is open). It is realized that it is necessary to use the EMERGENCY CARGO OPEN switch, and when that action is taken, the cargo door latch is indeed released, permitting the door to be opened. But at the same time, the landing gear is lowered, not a desired action and one that will probably aggravate the emergency. The condition that permits this undesired lowering of the landing gear to occur when both cargo door switches are closed is a sneak circuit.

The following two observations generally apply about sneak circuits:

1. Switches or other control elements are operated in an unusual or even prohibited manner.
2. The unintended function (in this example, the lowering of the landing gear) is associated with current flow through a circuit element that is opposite to the intended current flow.

The latter of these conditions permits elimination of the sneak circuit by inserting a diode, as shown in Figure 4.10. It is also the basis of an SCA technique described in Section 4.2.2.

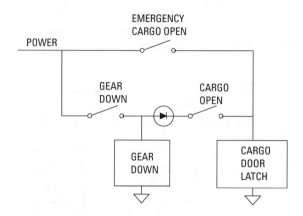

Figure 4.10 Corrected cargo door latching circuit.

Conventional SCA techniques depend on recognition of circuit patterns or "clues" to detect potential sneak circuits. The most common of these circuit patterns are shown in Figure 4.11.

The rectangular symbols represent arbitrary circuit elements, not exclusively resistors. In many cases the individual legs of the patterns include switches. It will be recognized that the leg containing the normal CARGO OPEN switch in Figure 4.9 constitutes the middle horizontal leg of an H-pattern. The

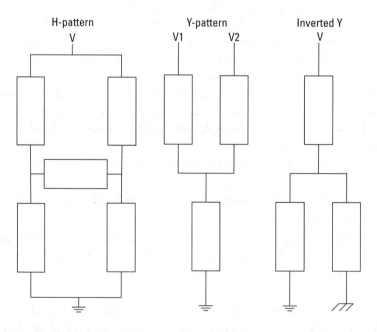

Figure 4.11 Circuit patterns for SCA.

Y-pattern is also called a power dome. The two upper legs terminate at different power sources, such as V1 and V2. The inverted Y is also called a ground dome; note that the two bottom legs terminate in different ground levels, such as chassis ground and signal ground.

To facilitate the recognition of these patterns or clues, the schematic diagrams had to be redrawn as "network trees," with power sources at the top and grounds at the bottom. In SCA, both positive and negative power sources will be shown at the top of the figure. Because searching for the patterns is very labor intensive, computer programs were developed to recognize the common clues in the network trees. Even with the aid of computers, SCA remained a costly and lengthy activity, and it was usually conducted only after the circuit design was frozen to avoid having to repeat it after changes. This had a distinct disadvantage when a sneak circuit was detected: it became very expensive to fix it because usually the circuit card or cabling was already in production.

4.2.2 Current SCA Techniques

As we mentioned in connection with the cargo door in Figure 4.9, sneak circuits involve current flow in an unexpected direction. This can also be verified in Figure 4.11. The normal direction of current flow is top down (and in the H-pattern from left to right). But under some conditions current can also flow from V1 to V2 in the Y-pattern, or from chassis to signal ground in the inverted-Y. If designers do not anticipate this reverse current flow, it can cause undesirable actions. Current SCA computer programs therefore identify paths that permit reverse current flow (also called bipaths), and this is considerably simpler than the earlier pattern recognition approach. It also eliminates the need to construct network trees.

Whereas the earlier SCA techniques were aimed primarily at equipment that used switches and relays, current designs use solid state devices to accomplish the same functions. The previously mentioned techniques can still be applied by using equivalent conductive path symbols for the semiconductors. Other changes in the conduct of SCA have been due to miniaturization and integration of computational and instrumentation circuits with the switching of sensitive functions. This has made it necessary to edit the circuit diagrams prior to conduct of the SCA.

An example of the integration of these functions for a hypothetical missile detonation system is shown in Figure 4.12. The computational elements at the top of the figure establish the conditions for operation of the prearm, arm, and detonate switches. The built-in test (BIT) legend on some paths refers to BIT signals used to monitor equipment operation. The heavy lines constitute the switching elements. The circles represent electroexplosive devices (squibs). The switching path is designed to prevent unintended detonation unless all three

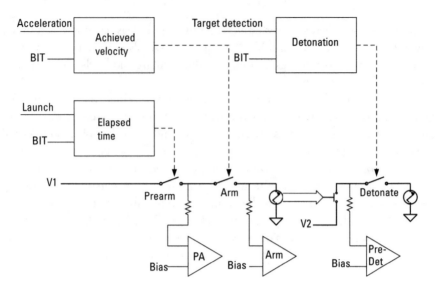

Figure 4.12 Hypothetical missile detonation system.

computational functions cause untimely switch closures. The instrumentation functions, shown in the lower part of the figure, furnish outputs for telemetry. SCA encompasses only the switching functions; the computational and instrumentation elements are eliminated from the traced paths.

This editing is justified because the connection between the computational elements and the switches (shown as dashed lines in the figure) is nonconducting. In most cases the output of the computational element goes to the gate of a Metal Oxide Semiconductor Field Effect Transistor (MOSFET), while the switching function uses the source-drain path. The computational elements are typically quite complex and their failure probability is much higher than that of the switching path. Thus the switching path is designed to tolerate the worst failure modes of these devices, and an SCA of the computational elements is not required.

The elimination of the instrumentation functions is justified by the isolation resistors at the connection with the switching function. The resistance values are typically on the order of 10k ohms. Since the switching voltage is in the 20- to 30-V range, the current flow through the isolation resistors cannot exceed a few milliamperes, while squibs fire only above 1A.

With the techniques described in this subsection, SCA of even complex systems can be accomplished at a reasonable cost.

4.3 Fault Tree Analysis

Fault tree analysis (FTA) complements FMEA by starting with a top-level failure effect and tracing the failure to potential causes. It is not restricted to the effects of a single failure, and indeed is aimed at detecting how multiple lower level events can combine to produce an undesirable top-level effect. FTA is an important component of reliability analysis and safety programs. The following describes the basic concepts of FTA and discusses an example.

4.3.1 Basics of FTA

Whereas FMEA encompasses all parts or functions of a component, FTA is applied selectively to the most severe failure effects. A first step in conducting an FTA is therefore to identify the effects (or events, as they are usually referred to in this context) to be analyzed. Listing category I failure effects from the FMEA can serve as a starting point. However, in a well-designed component or system no single failure should give rise to a critical failure, and other sources, such as a hazards analysis, may have to be consulted. For most electrical household appliances there are two top events for which FTA may have to be generated: electric shock and fire. For a passenger elevator there may be so many top events that each have to be analyzed separately and in combination (e.g., simultaneous fire and inability to move represents a more serious hazard than either condition by itself).

Once a top event has been selected, the analyst must identify its immediate causes. These can be additive (either cause A or cause B will result in the top event) or complementary (both must occur to cause the top event). Additive causes are usually represented by an OR gate, and complementary causes are represented by an AND gate. These and other symbols used in FTA are shown in Figure 4.13.

The exclusive OR (XOR) is used when either of two events, but not both together, can cause an upper level event. This condition is encountered in three-way switching circuits, where one switch can energize a device but subsequent operation of the other switch will restore it to its previous condition. The circle symbol is a basic failure event that will not be expanded further. The FMEA ID of the failure mode can be inscribed in the circle. The probability of failure is usually denoted under the circle. The continuation symbol is used to permit complex fault trees to be shown on more than a single page. FTA is a

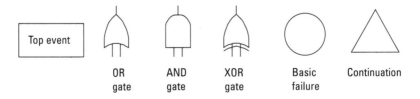

Figure 4.13 FTA symbols.

hierarchical procedure, and failure probabilities calculated from lower level analysis on one page can be propagated to the higher levels. The use of the symbols is illustrated by an example in Section 4.3.2. Only failures in electrical or electronic components are treated there. In many situations fault trees must consider combinations of anomalies in electrical, structural, and thermal environments. For nuclear power plants and many other utility services, additional hazards due to weather, earthquakes, and acts of war have to be included.

4.3.2 Example of FTA

This example uses the missile detonation diagram first introduced in Figure 4.12. The fault tree is applicable to the flight phase, and it assumes that V1 and V2 are present. A similar tree for the preflight phase will make different assumptions. The top event to be investigated is premature detonation, defined as firing prior to intended issuance of the arm signal (detonation after arm but prior to issuance to the detonation signal is undesirable but not unsafe). For the purpose of this FTA, the representation of the circuit has been modified, as shown in Figure 4.14.

The modifications are the following:

- Labeling of switches S1–S4 for reference in the fault tree;
- Replacement of sensed quantities at the input to computing blocks with the generic term "sensor" to facilitate construction of a common subtree for all three computing blocks;
- Deletion of the instrumentation functions.

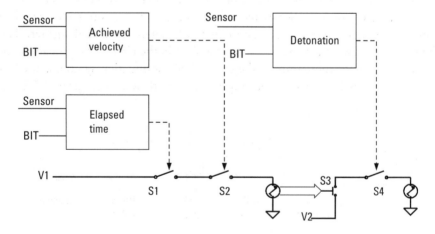

Figure 4.14 Missile detonation circuit modified for fault tree.

The last modification is justified by the low current output of the instrumentation power supplies, typically at least 1,000 times below the level required for squib detonation, and by the presence of isolation resistors. However, the instrumentation circuits can detect potentially hazardous conditions before they cause the top event, and that capability may be taken into account in the FTA of an actual system. It is not essential for our demonstration of the principles of FTA.

As previously mentioned, we want to take advantage of the identical structure of the three computing blocks by constructing a subtree, as shown at the lower right in Figure 4.15 (continuation triangle 1). The numbers shown in each basic failure circle represent mission failure probabilities. Each computing

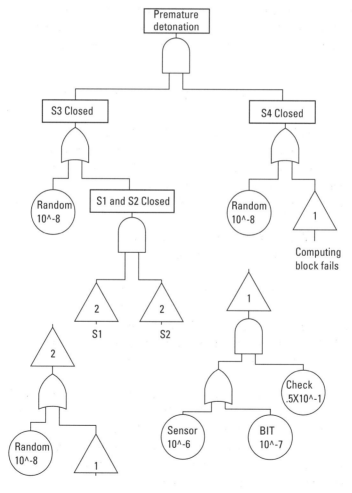

Figure 4.15 Fault tree of missile detonation circuit.

block receives two inputs, sensor and BIT, and faulty signals entering by either of these paths may cause a faulty output. This condition is represented by the OR gate. Because faulty inputs must be anticipated, the computer program protects against this by check routines (also called reasonableness tests) that are here assumed to be 0.95 effective in detecting errors (0.05 check failures). The failure probability associated with subtree 1 can now be computed as

$$F_{S1} = \left(10^{-6} + 10^{-7}\right) \times 0.05 = 0.055 \times 10^{-6} \qquad (4.2)$$

where the quantity in parentheses is the sum of the sensor and BIT failure probabilities.

Another equivalence is present in the operation of switches S1 and S2; this is represented by subtree 2 (shown in the lower left of Figure 4.15). Each of these switches can fail due to a random event or because of a failure in the associated computing block (subtree 1). The failure probability for subtree 2 can be computed as

$$F_{S2} = 10^{-8} + 0.055 \times 10^{-6} = 0.065 \times 10^{-6} \qquad (4.3)$$

The probability of the event "S1 & S2 Closed" in the main tree is the square of F_{S2}; it can be neglected because it is many orders of magnitude below the other input to the "S3 Closed" OR gate. The probability of "S3 Closed" event is therefore the random failure that carries a probability of 10^{-8}. The structure of the events leading to "S4 Closed" is seen to be identical to that of subtree 2, $F_{S2} = 0.055 \times 10^{-6}$.

The probability of premature detonation, F, is the product of the probabilities of "S3 Closed" and "S4 Closed."

$$F = 10^{-8} \times 0.055 \times 10^{-6} = 5.5 \times 10^{-16} \qquad (4.4)$$

For civil transport aircraft, failure probabilities of 10^{-9} or less are prescribed [13]. For military applications, higher failure probabilities are sometimes acceptable. By any measure, the FTA of the detonation circuit shows that it will meet requirements.

This analysis assumes that the failure of all computing blocks and of switches S1 and S2 are independent. If these conditions are not met (e.g., because all computing blocks are part of one chip), the probability of correlated failures must be taken into account, and this can be the governing failure probability. Also, the only failures in the computational blocks analyzed were those due to inputs and processing. Total computer failure is easily detectable and will not cause premature detonation.

This example has been concerned only with electrical failures. Particularly when it supports safety concerns, FTA may have to include all causes of failure, including those due to mechanical, environmental, and social conditions (acts of war or sabotage).

4.4 Chapter Summary

Analytical techniques usually provide insight into defined causes of unreliability, and they can be applied early in the development when many corrective measures can be brought to bear. They are usually not effective against failures due to system or operator interactions (such as the organizational causes in Chapter 3), unexpected environmental conditions, or errors or omissions in the specification. The analytical techniques need to be validated for the intended application. A specific example of the latter concept is that failure modes and effects used in the FMEA should be validated against the failures experienced by predecessor products and should continue to be validated against failures in operation. This latter technique is discussed in Chapter 8.

References

[1] Cunningham, T. J., and Greene, K., "Failure Mode, Effects and Criticality Analysis," *Proceedings of the 1968 Annual Symposium on Reliability,* January 1968, pp. 374–384.

[2] McCready, K. F., and Conklin, D. E., "Improved Equipment Reliability Through a Comprehensive Electron Tube Surveillance Program," *Proceedings of the First National Symposium on Reliability and Quality Control,* November 1954, pp. 81–94.

[3] Automotive Industry Action Group (AIAG), "Potential Failure Modes and Effects Analysis" (technical equivalent of SAE J-1739), Third edition, July 2001.

[4] Goble, W. M., *Control System Safety Evaluation and Reliability,* Second edition, Research Triangle Park, NC: ISA, 1998.

[5] Department of Defense, "Procedures for Performing a Failure Modes, Effects and Criticality Analysis," AMSC N3074, November 24, 1980.

[6] *RAM Commander with FMEA,* SoHaR Inc., Beverly Hills, CA, 2002.

[7] *Relex FMEA,* Relex Software Corporation, Greensburg, PA, 2002.

[8] Goddard, P. L., "Software FMEA Techniques," *Proc. of the 2000 Reliability and Maintainability Symposium,* Los Angeles, CA, January 2000, pp. 118–122.

[9] *Specification for UML Vers. 1.4,* Object Management Group (OMG), Needham, MA, 2002.

[10] Douglass, B. P., *Real-Time UML,* Second edition, Boston, MA: Addison Wesley, 1999.

[11] Rankin, J. P., and White, C. F., "Sneak Circuit Analysis Handbook," 1970, National Technical Information Service (NTIS N71-12487).

[12] U.S. Navy, "Contract and Management Guide for Sneak Circuit Analysis," 1980, (NAVSEA-TE001-AA-GYD-010/SCA).

[13] Federal Aviation Administration, Advisory Circular 25.1309-1A, "System Design and Analysis," June 21, 1988, U.S. Department of Transportation.

5
Testing to Prevent Failures

Testing is a costly activity, but it is essential for the reliability assessment and acceptance into service of critical components and systems. A number of statistical techniques exist to demonstrate that the MTBF is not less than a specified number, and these are briefly described in this chapter's first section. The techniques require that a large number of articles be put on test at the same time (or that a long test time be scheduled), and they are therefore more suitable for parts and low-level components rather than at the system level.

An alternative approach is to use testing to verify or establish design margins and to transform these into a probability of failure, as discussed in the chapter's second section.

To select tests that are practical for assessing the reliability of a complex system, we survey the overall spectrum of tests (not restricted to those for reliability) that a system and major components can be expected to undergo. First we cover tests during development in Section 5.3 and then postdevelopment tests in Section 5.4. The importance of in-service tests (such as self-test and BIT) is steadily increasing, and the reliability implications of in-service testing are discussed in Section 5.5. Section 5.6 summarizes the chapter.

Before delving into the details of testing to prevent failures, the following apply to all types and levels of tests:

1. Early testing permits the results to be brought to bear on the design and development; problems found in late testing are much more difficult to correct and are frequently accepted as a permanent limitation of the system.

2. Before any testing for contractual purposes is undertaken, there should be a clear understanding of what constitutes a "chargeable failure" and whether failed articles should be replaced during the test. Conflicts can arise when failures occur shortly after a power interruption (they may be due to high transient voltages), when operators interrupt the test to troubleshoot test equipment, or when the random failure of one item may cause overheating of adjacent ones.
3. In reliability testing, it is usually not a question of accepting or rejecting a production lot; rather it is to find and correct causes of failure. For that reason, a failure analysis and corrective action procedure (FRACAS) should be in place before testing is started. FRACAS is discussed in Section 5.3.

5.1 Reliability Demonstration

In principle, the concept of reliability demonstration is very simple: a representative large number of the product is put on life test under conditions equivalent to the usage environment. When an item fails, the time of failure is recorded and the test is continued until every item under test fails. The failure times are then totaled and the sum is divided by the population size. The result is the demonstrated MTBF. If the product is reasonably reliable it could take very long to complete this test, and it is impossible to know in advance how long.

To overcome this serious problem, procedures have been devised to terminate the test either at a specified interval or after the specified number of failures have been observed. Let us look at the fixed interval methodology and apply it to a transistor for which the manufacturer claims an MTBF of 10^6 hours. The transistor is assumed to fail in accordance with the exponential failure law under which the failure probability is independent of the operating time. Therefore it makes no difference whether we test one transistor for 10^6 hours or 1,000 transistors for 1,000 hours. Something closer to the latter alternative will generally be preferred, and the following discussion is based on that format.

The 1,000 transistors are put on test, and a failure is recorded at 500 hours (the failed transistor was not replaced) and a second one at 999 hours. The vendor claims that the failure at 999 hours was so close to the test termination that, in effect, there was just one failure in the almost 10^6 hours of testing and the claim of 10^6 hours MTBF has been substantiated. The customer retorts that you cannot be sure that there would not be additional failures if the test had been continued for even a few hours more and that the MTBF claim had not been substantiated. This example illustrates the fundamental problem of *truncated* reliability demonstrations, but we have seen earlier that it is practically impossible to run nontruncated tests. The concepts of producer and consumer risk that were extensively analyzed between 1970–1980 [1] deal with this problem. The following conceptual explanation may suffice for most system reliability evaluations.

In the previous problem, the failure rate, λ, was 10^{-6} and the failure exposure time, t, was 10^6 hours. Thus, the product $\lambda t = 1$ and the expected probability of failure is 0.63 [see (2.3)]. From the Poisson probability law, the probability of x failures is shown in Table 5.1.

One interpretation of this table is that two or more failures can be expected with probability of 0.2642 (the sum of the last four entries). Rejection of the MTBF claim when two or more failures have been observed thus subjects the producer to a 0.2642 risk of having the product rejected even if it had indeed a true MTBF of 10^6 hours.

Assessment of the consumer's risk must be approached somewhat differently. Is the item really unusable if the MTBF were 0.9×10^6? Probably not. If it were 0.09×10^6? The ratio of the specified MTBF, θ_0, to the lowest one that is definitely not acceptable to the consumer, θ_1, is called the *discrimination ratio*, usually designated Δ. Frequently used values for this ratio range from 1.5 to 3, and the continuation of the example uses $\Delta = 2$. Thus, the consumer's risk is that the actual MTBF is 0.5×10^6 or less when two failures are observed in 10^6 hours of testing. The λt product for these conditions is 2, and Table 5.2 shows the corresponding failure probabilities.

It is seen that even with the acceptance criterion set at one failure in 10^6 hours, the consumer faces a risk of 0.4060 that the actual MTBF is less than one-half of the specified value. To reduce both consumer and producer risk it is necessary to increase the test duration. The effect of increased test duration can be seen in Table 5.3, which is excerpted from Table 5.1 of [2]. The consumer and producer risk are each at the value listed in the risk column. The test duration is expressed in multiples of the specified MTBF, θ_0.

Table 5.1
Probability of Failure for $\lambda t = 1$

x	P(x)
0	0.3679
1	0.3679
2	0.1839
3	0.0613
4	0.0153
≥5	0.0037

Table 5.2
Probability of Failure for $\lambda t = 2$

x	P(x)
0	0.1353
1	0.2707
2	0.2707
3	0.1804
4	0.0902
≥5	0.0527

Table 5.3
Effect of Test Duration on Risk

Δ	Risk	Duration
1.5	0.1	30
1.5	0.2	14.3
2	0.1	9.4
2	0.2	3.9
3	0.1	3.1
3	0.2	1.4

The major reason for undertaking a reliability demonstration is to have a contractually enforceable mechanism for rejecting a product when its reliability falls short of requirements. This benefit is obtained at considerable cost and loss of "time to market," and it is also contingent upon the consumer having an alternative in case of a reject decision. This condition is seldom met in the procurement of critical systems. We will therefore now look at other methods for assessing the reliability of a system or its major components.

5.2 Design Margins

As a product is being developed, designers and component specialists accumulate data that has important bearing on its reliability. This section discusses the importance of tests that establish design margins for a product's reliability.

Mechanical and structural engineers had long used safety factors, based on craft traditions, to prevent failures in service. During the early part of the twentieth century, there was increasing interest in replacing traditional practices with rational design procedures. A significant breakthrough in these efforts was the assessment of the strength of cables for the Golden Gate Bridge by Freudenthal [3]. He examined test records and found that the tensile strength of the cables had a Gaussian distribution and therefore the knowledge of the mean strength and standard deviation permitted calculation of the probability of failure under any given value of load. This relationship is illustrated in Figure 5.1. Because the strength should always be greater than the load, the margin has a negative value. Note the logarithmic scale for the ordinate. The normalized margin, M, is computed from

$$M = (L - S) / \sigma_S \tag{5.1}$$

where L = Load (assumed to be a fixed value);
S = Strength (mean value, same units as load);
σ_S = Standard deviation of strength.

Exact values are difficult to determine from the figure and are therefore listed in Table 5.4.

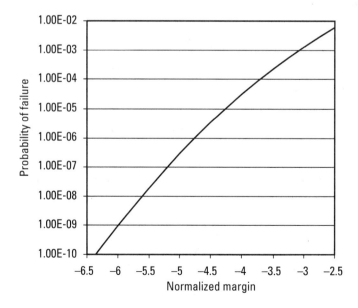

Figure 5.1 Probability of failure versus normalized margin.

Table 5.4
Probability of Failure for Low Values of M

M	Probability of Failure
−6.5	4.03579E-11
−6	9.90122E-10
−5.5	1.90364E-08
−5	2.87105E-07
−4.5	3.40080E-06

Assume that a cable has a mean strength of 10,000 kg with a standard deviation of 1,000 kg. A load of 4,000 kg will therefore produce $M = -6$, and the probability of failure indicated in Figure 5.1 is approximately 10^{-9}, a value generally accepted as "extremely unlikely." The corresponding safety factor (S/L) of 2.5 was commonly used for the highest quality construction practices. In that respect, the two procedures are in agreement. But the safety margin approach showed that if tighter manufacturing controls could reduce the standard deviation to 500 kg, the same normalized margin would then correspond to a load of 7,000 kg. The advantages of the safety margin approach did not take long to be widely recognized and were put to practice in statistical quality control. Reliability engineers found that not only structural properties but also electrical characteristics could be analyzed by design margin.

One typical application is in estimating the failure probability of a radar receiver that consists of two stages, as shown in Figure 5.2. A threshold circuit in the detector (intended to reject noise) allows an output only for signals exceeding the threshold. The amplifier was designed to provide an amplified signal that was in excess of the threshold whenever a valid input was received. A number of unexplained malfunctions were reported that could not be traced to a physical component failure. The investigation showed that the amplifier gain, as well as the detector threshold, had wide tolerances, such that a low gain amplifier would fail to furnish enough signal to drive a high threshold detector.

The statistical parameters of conditions leading to undetected signals are shown in Figure 5.3, where the solid line represents the cumulative probability distribution of the amplified signal while the dotted graph is the probability density function (pdf) of the detector threshold. The mean of the detector threshold is two standard deviations below the mean amplifier output, and thus in most cases a valid signal will be detected. But, as seen in the figure, there is

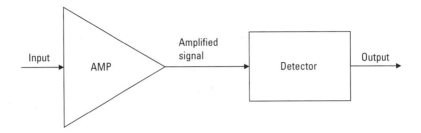

Figure 5.2 Radar receiver.

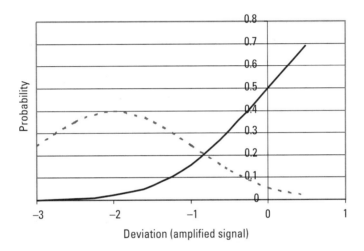

Figure 5.3 Probability of undetected signal.

approximately 0.02 probability that the amplified signal will be at or below the mean detector threshold, and much higher probabilities that it will be below the upper end of the threshold distribution. The failure probability can be calculated by means of a convolution integral in which each increment of the threshold pdf is multiplied by the value of the amplified signal distribution at the same point along the *x*-axis.

Where the distributions are normal, a much simpler approach is possible by working with the difference, $D = A - T$, that has a normal distribution with mean and standard deviation given by

$$m_D = m_A - m_T \qquad \sigma_D^2 = \sigma_A^2 + \sigma_T^2 \qquad (5.2)$$

where the subscripts *D*, *A*, and *T* stand for the difference, amplified signal, and detector threshold distributions, respectively.

The probability of failure is equivalent to the probability that the difference distribution will have a negative value. For the above example $m_A = 0$, $m_T = -2$, and $m_D = 2$. Also, $\sigma_A = \sigma_T = 1$, $\sigma_D^2 = 2$, $\sigma_D = 1.41$. The mean m_D is located $2/1.41. = 1.41$ standard deviations above zero. The probability of the variate $(A-T)$ having a negative value is approximately 0.08.

When the methodology of this section is compared to that of the reliability demonstration in Section 5.1, two significant differences emerge:

1. In Section 5.1, the test outcomes were characterized at success or failure; this is called *testing by attributes*. The kind of failure and the extent of deviation from the specification are not considered. The design margin approach of the current section takes note of the numerical test outcomes, not only in case of a failure but for every article tested. This is called *testing by variables*.
2. The reliability demonstration covered all causes of failure. The design margin approach covers only specific causes for which data is being analyzed. Thus, the investigation of the amplifier in the last example does not tell us anything about failures due to lead breakage or solder balls.
3. Testing by variables provides an opportunity to reassess the design margin at later time intervals or after some critical events have occurred. In the last example, retesting the amplifier gain after the first year of operation can establish quantitatively whether the probability of failure has increased.

In general, testing by variables provides more insight into causes of failure and therefore their prevention. The reliability assessment is not as contractually effective as that of reliability demonstration but can be obtained in much less time and with fewer test specimens. It is sometimes held that digital circuits can only be evaluated by attributes. A counter or an adder either works or does not, and their performance does not indicate probability of failure. But these circuits can be operated at lower-than-specified voltage or higher-than-specified clock rates to establish design margins. In Section 5.3, we look for opportunities to glean indicators of reliability from other test activities that are usually conducted during the development cycle.

5.3 Reliability Relevance of Tests During Development

It bears repeating: Testing is a very expensive activity. Good test preparation and documentation (see Insert 5.1) are essential. From the reliability perspective, a good FRACAS is even more important. A good FRACAS turns every failure into a learning opportunity. The following are key entries in the FRACAS:

> **Insert 5.1—Significant Test Documents**
>
> Test Plan
>
>> The test plan covers the entire development phase and describes the purpose, locale, date, facilities, and responsible organization for each test. Revisions of the test plan contain successively more detail about each test.
>
> Test Specification
>
>> A separate test specification is prepared for each test. The specification describes the units under test (UUT), the characteristics to be tested, how each measurement will be recorded, the test fixtures to be used, and how test results will be reviewed. The test specification is usually prepared by the developing organization.
>
> Test Procedure
>
>> The test procedure translates the requirements of the test specification into detailed actions to be taken at each step in the test. Where the test specification may say "measure output at zero, one-half and full input signal," the procedure will say "Connect typeX903 output meter to terminals 516 and 517. Set switch S5 to 0. Let output meter stabilize for at least 5 seconds and record reading. Repeat with switch S5 at settings of 5 and 10." The test procedure is usually prepared by the test group.
>
> Test Report
>
>> The test report is prepared at the conclusion of each test. An executive summary usually provides highlights, including details of any failures that were encountered. Subsequent sections describe the equipment that was tested, the test environment, and the sequence of test events. Detailed test results are keyed to paragraphs of the test specification and may be presented in the body of the report or in an appendix.

1. Date and time of incident;
2. Location and special environmental circumstances (rain, extreme cold);
3. Operation in progress at time of failure;
4. First observation of failure (system abort, wrong output, alarm) and by whom;
5. Confirmation of failure (restart, special diagnostic, replacement) and by whom;

6. Type and serial number of removed component(s);
7. Time of resumption of operation;
8. Later findings on component(s) responsible for the failure;
9. Disposition of failed component(s);
10. Failure mode and effects coverage of this failure.

From this information, the Failure Review Board (FRB) may make one of the following determinations:

1. Failure not verified (FNV) or could not duplicate (CND). The affected component is usually returned to service.
2. Failure was a random event. The component is repaired and returned to service.
3. Failure was not covered by FMEA; revise FMEA and consider classification 5. This finding is an important feedback on the quality of the FMEA. If many failures are found that were not listed in the FMEA, the methodology for generating it may have to be revised.
4. Failure was induced due to use outside the specified environment or failure of another component (e.g., overheating because of fan failure). The component is repaired and returned to service.
5. Failure was due to design or requirements deficiency. Requirements and/or design must be modified to prevent recurrence of this failure mode. The component is returned to service only after the design change has been incorporated. If the failure can produce safety or mission-critical effects, all components of this type will have to be modified. In the mean time, the system cannot be operated at all or must be placarded (restricted to operate only under conditions that reduce the likelihood of this failure).

The importance of the failure reporting system is that over time, many failures in other classifications can be reclassified into category 5 and made the subject of a design change that results in a reliability improvement.

The FRACAS and the test documents shown in the text box are the infrastructure for all tests. Without that infrastructure, the resources devoted to testing are largely wasted. A listing of typical tests during the development phase is shown in Table 5.5.

Further details of these tests and particularly their reliability implications are presented in the following sections.

Parts Evaluation

When a design calls for a new part or subcomponent, such as a power supply, the engineering group usually purchases a few samples to test them on the

Table 5.5
Typical Tests and Their Purpose

Type of Test	Primary Purpose
Parts evaluation	Establish suitability of (new) parts for the intended use
Breadboard test	Verify performance against development goals
Component prototype test	Verify performance and attributes against specification
System integration test	Determine compliance with system specifications
Qualification test	Verify performance at extremes of operating environment
First article acceptance	Determine compliance with specification
Operational evaluation	Determine suitability for the intended operational use

intended circuit or mechanical assembly. Quantitative recording of the test results can answer the following questions:

- Does the mean of the principal performance or dimensional parameters correspond to the published nominal?
- Is the standard deviation in the parameters consistent with the published performance or dimensional tolerances?
- Which circuit or assembly characteristics are particularly sensitive to the variability of the part that is being evaluated? Can larger-than-published tolerances be accommodated?

Breadboard Test

Usually only one breadboard is being built and therefore unit-to-unit variability cannot be assessed in this test. However, the following observations can lead to possible causes of failure:

- Were any changes in parts necessary to achieve the specified performance? If so, were the reasons for this understood and were design practices adjusted to help in future designs?
- Was the achieved performance in the mid-range of the expectations?
- Were transients during start-up, shutdown, and mode changes within the expected envelope?
- Was performance checked at high and low limits of supply voltage and clock frequency?

Component Prototype Test

The following are key reliability-related questions that arise in the prototype test:

- Does the prototype performance match that of the breadboard? If not, are the reasons for this understood?
- Is the unit-to-unit variability consistent with the specified output tolerance?
- Are there unexpected local hot spots or unexpected friction or play?
- Have displays and controls been used by personnel representative of the operator population?
- Has accessibility of replaceable parts been reviewed by personnel representative of the maintenance population?

System Integration Test

This is usually the first opportunity to observe the operation of a complete system, and many reliability-related issues can be evaluated. Among these are the following:

- Were transients during start-up, shut-down, and mode changes within the expected envelope?
- Was there a correct response to power interruptions?
- Were any changes in components necessary to achieve the specified performance? If so, were the reasons for this understood and were design practices adjusted to help in future designs?
- Was the achieved performance in the mid-range of expectations?
- Was the response to faulty input signals or operator inputs as expected?
- Did self-test provisions detect component faults?

Qualification Test

The qualification test subjects the equipment to environmental conditions that are intended to simulate the extremes that are expected in service. Typical conditions are combined temperature, humidity, and vibration, as well as separate tests for shock, exposure to high magnetic and electric fields, nuclear radiation, salt spray, and specialized tests depending on system characteristics (e.g., centrifuge or rocket sled to generate high accelerations). The test thus requires extensive test chambers and test fixtures. It is rare if equipment passes these tests without failures and typically these are classified as design deficiencies that must be removed before production is authorized.

In addition to the obvious reliability concerns with observed failures, the test data may be evaluated for the following issues:

- Are there permanent changes in performance as a result of the environmental exposure? Do these need to be evaluated for in-service significance?
- Is the observed performance variation at environmental extremes as expected?
- Were output transients observed in vibration, shock, and other tests? Are these operationally significant? Do they predict permanent outages in prolonged exposure to these environments?

First Article Acceptance Test

All the preceding tests are usually run with units that were built in a prototype shop staffed by very skilled personnel under direct supervision of the developers. The first article acceptance test is intended to establish that units built with production fixtures on the factory floor furnish equivalent performance. The first article acceptance test is much more thorough (involves more characteristics, input values, and operating modes) than the routine acceptance test. Reliability concerns that may be evaluated include the following:

- Stability of performance parameters throughout the test;
- Variability of performance parameters if several units are tested;
- Performance of all self-test and redundancy provisions.

Operational Evaluation

Important systems are carefully monitored when they are first put into operation. The military services use the term "OpEval" for this activity and usually follow a prepared script for it. Other organizations call it initial operational capability (IOC) and provide for monitoring and technical support by the developer but otherwise let operating personnel follow their normal routine. Human interface and maintainability concerns are more frequently encountered than reliability problems. Circumstances that may give rise to the latter include the following:

- Noisy power and signal lines;
- Electromagnetic radiation in the environment;
- Interruption of air conditioning and heating (e.g., over weekends);
- Abuse of the equipment in shipping and installation.

The contractual responsibility for correcting these problems varies from case to case but the failure mechanisms encountered in the operational evaluation are likely to be representative of the general user environment and must therefore be addressed.

5.4 Reliability Relevance of Postdevelopment Tests

During the development phase, tests are extensive and conducted on only one or a few units. In contrast, the typical postdevelopment test is much less extensive but conducted on many units. This provides an opportunity to use methods of statistical quality control to gain insight into trends that may affect reliability. Typical tests in the postdevelopment period include the following:

- Routine acceptance tests;
- Screening tests;
- Field failure investigations.

Some organizations conduct periodic intensive acceptance tests (e.g., on every hundredth unit). The reliability implications of these tests are essentially the same as those for routine acceptance tests. Failures encountered in any of the postproduction tests should be subjected to the same failure reporting requirements as those discussed for the development tests.

Routine Acceptance Tests

These tests capture parameters that are of interest to the user. We focus on whether the data that is collected from acceptance tests can be used to identify potential causes of failure. The first representation of acceptance test data that will be examined is in the form of a control chart. as shown in Figure 5.4.

The graph shows the value of output (under specified input conditions) for 100 units in the order in which they were tested. The minimum acceptable value for the output is 5.5 units (indicated by the dotted line), and there is no upper limit. Out of these 100 units, none were rejected for low output but the graph has some characteristics that should be investigated for reliability implications. The first of these is the cyclic pattern evident for the first 40 units and then again for the last 50. Such patterns are sometimes due to a shift effect; assemblers first take some time to warm up and later on in the shift they may get tired. If a parameter is sensitive to assembly skill, it can be expected to exhibit this pattern. In this instance, units 65 through 80 seem to present reliability concerns because of low output, and the circumstances under which they were assembled and tested should be investigated.

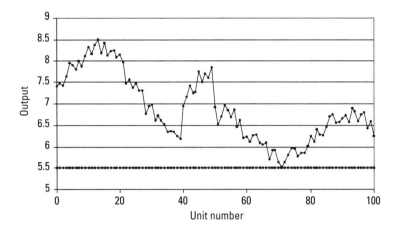

Figure 5.4 Control chart.

The second observation concerns the abrupt increase in output starting at about unit 40 and the abrupt decrease in output starting at about unit 50. This is not a random pattern, and the cause of it must be understood. A final observation is that the average output of the last 50 units is well below that of the first 40. Units 65–80, for which reliability concerns were previously indicated, are included in the low output group. Other uses of control charts are discussed in Section 8.5.

Another way of looking at this data is to arrange it in the order of increasing output measurement, as shown in Figure 5.5. Such an ordering is frequently called a Pareto chart, after Italian economist Vilfredo Pareto (1848–1923), who used this presentation to show that a small fraction of the population controlled

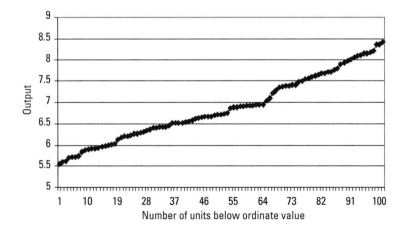

Figure 5.5 Pareto distribution.

the majority of the wealth. For our purposes, the figure shows that about 15 units have output values below 6, and that there are several discontinuities centered on an output value of 7. Both observations can be correlated with Figure 5.4.

Yet another cut at the same data is presented in the frequency distribution shown in Figure 5.6. It illustrates more clearly than the others that there are 16 units with output values below 6, and it exhibits a noticeable deviation from the normal distribution, particularly the dip in the 7–7.49 output range (associated with the discontinuities previously discussed). Such deviations from the normal distribution frequently indicate that portions of the manufacturing and assembly process are not well controlled and thus are a potential source of reliability problems.

Routine acceptance tests are not intended to demonstrate reliability. But areas of concern could be identified in these examples, and if such deviations are encountered in practice they should be investigated.

Screening Tests

As the name implies, these tests screen by examining the values of a number of parameters, and that number must be kept within bounds to make the tests economically viable. For this reason, screening tests are usually conducted at the part or assembly level where performance can be characterized by a reasonable number of measurements. Screening can be performed in a quiescent environment with the aim of eliminating units at or below minimum performance levels, or in a stressed environment (stress screening) with the aim of eliminating units that show excessive parameter variation in response to environmental

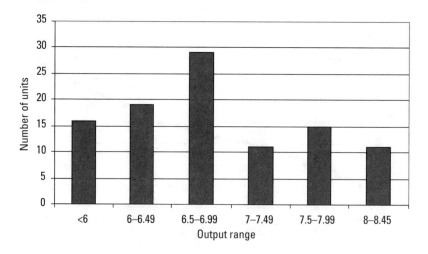

Figure 5.6 Frequency distribution of output range.

change. We discuss primarily the former; the methodology for stress screening is essentially the same but the interpretation of data is more application-dependent.

For example, a manufacturer of cordless phones has received returns at a 5% rate because of fading and eventually complete loss of reception. The investigation points to high leakage current in the output filter capacitor of the power supply. Initially this causes lowering of the output voltage and it can progress to a complete short circuit. Space restrictions prevent use of a capacitor with higher voltage rating. From tests on several lots, it is found that the leakage current has an approximately normal distribution with mean of 10 μA and a standard deviation of 1 μA. The inverse cumulative distribution corresponding to these measurements is shown on Figure 5.7.

Since 5% of the phones were returned, it might be assumed that eliminating 5% of the capacitors with the highest leakage current will solve the problem. The corresponding rejection criterion is seen in Figure 5.7 to be approximately 11.8 μA. To be on the safe side, it was decided to screen out all capacitors with a leakage current of 11 μA or greater, leading to rejection of approximately 16% of the product. The resulting frequency distribution of the screened capacitors is shown in Figure 5.8.

This distribution is no longer normal, and a large proportion is just below the rejection point. But since the capacitors responsible for the phone failures appear to be ones with more than 11.8-μA leakage current, the screening criterion of 11-μA may be an acceptable business decision.

An important use of screening is to reduce failures occurring during the initial period of use, frequently referred to as infant mortality. Moderately

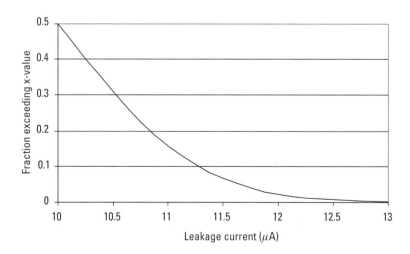

Figure 5.7 Leakage current of capacitors.

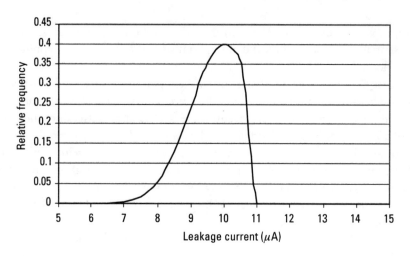

Figure 5.8 Frequency distribution after screening.

elevated environmental stresses are usually employed in this screening. The effect of infant mortality is depicted in the A–B segment of Figure 5.9. Most published failure rates apply to the B–C segment (constant failure rate), and thus infant mortality represents an increase above the expected failure probability. The segment to the right of C represents wearout and is not part of the current discussion.

The reliability function for the A–B segment is frequently modeled by the Weibull distribution as

$$R(t) = e^{-t^\beta/\alpha} \tag{5.3}$$

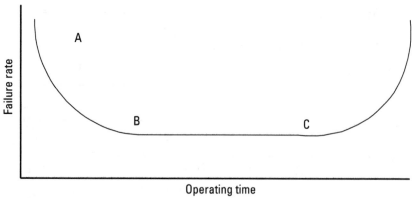

Figure 5.9 Infant mortality.

where the shape parameter $\beta < 1$. For $\beta = 1$ and $\alpha = 1/\lambda$ this becomes the exponential distribution. While the Weibull distribution is a useful empirical model, a mathematical formulation of infant mortality more closely tied to possible causes can be derived from statistics of extremes [4].

The cause of infant mortality is usually assumed to be defects in the manufacturing and assembly processes, and screening at stress levels moderately higher than those expected in use have been effective in reducing infant mortality experienced by the user [5]. The increased effectiveness of process control measures has also played an important part in preventing the defects to be introduced at all. It has also been suggested that the infancy period of the manufacturing process contributes to the early-life failures, in addition to product infancy [6].

Field Failure Investigations

In the previous example, failures occurred soon after the phones were put into use, and the cause of the failure could be immediately determined. At the other extreme are single failures that have catastrophic effects (see Chapter 3) and are therefore painstakingly analyzed by specialized teams. Here we address failures for which the cause is not obvious and that do not lead to serious accidents but do affect the acceptance of a product.

A good place to start is by classifying the reported failure events under headings such as the following:

- Time of day of the failure;
- Operation in progress at time of failure;
- Time since being put into service;
- Date of manufacture;
- Equipment location (hot, cold, wet, subject to power fluctuations, vibrating, and so on);
- User characteristics.

From these it is usually possible to establish test conditions that can be used to evaluate the potential for failure in the current product, to identify the design feature responsible for the failures, and to plan corrective measures.

If the symptoms are not complete failure but rather impediments to usage, like a flickering display or inability to enter into an operating mode, the development staff may have some clues regarding conditions that cause these malfunctions. Test programs can then be developed to explore the potential for failure and proceed as indicated in the previous paragraph.

Field failures can be caused by exceeding the operational or environmental envelope but unless there are direct indications that this was the cause, the manufacturer or vendor may still be morally or legally obligated to undertake the repair. For these reasons, it may be desirable to install temperature sensors that prevent operation at extreme heat or cold or to record maximum acceleration and to provide safeguards against improper operating sequences. A bundled training program can also be effective in reducing failures due to improper use of the equipment; this measure is most suitable for equipment sold to specialized markets.

5.5 In-Service Testing

This section deals with continuous monitoring by error detecting codes (EDC), monitoring at intervals with BIT, and periodic testing using a field test set (FTS).

Error Detecting Codes

EDCs can be used to assess whether digital information is correctly retrieved and transmitted [7, 8]. When an error is detected, the operation that gave rise to the error is sometimes repeated and this may overcome the failure. Alternatively, an error detecting and correcting code (EDAC) may be employed that can correct for a limited number of errors in one word or one transmission. If that fails, a duplicate storage may be accessed or retransmission of a binary string may be requested.

Whichever of these actions is taken, a record of error detection can be obtained. If error detection occurs infrequently, the measures discussed in the preceding paragraph may suffice. But if it is frequent, and particularly if there is an increasing frequency of error detection with respect to one memory unit or one transmission path, this becomes a reliability problem and calls for corrective action. Recording error detection activity can therefore be an important adjunct to the other postdevelopment testing that has already been discussed. Typical characteristics of the other two modalities are compared in Table 5.6.

The table shows that the approaches are complementary, and frequently both are used. Beyond the differences shown, a common concern is how much needs to be tested. For example, the tabletop TV set either works or it does not. There may be a few failure modes that affect only the UHF band, but they do not reduce my use of the set if I am not aware of them while watching on the VHF band. However, if I am saving a file on a computer disk, I want to be aware of all circumstances that might prevent access to that file in the future. In general, the need for current information about equipment capability increases when:

- The equipment has multiple operating modes;

Table 5.6
Typical Characteristics of BIT and Field Test

Characteristic	Built-In Test	Field Test Set
Primary purpose	Assess current status	Test readiness for future service
Frequency of test	At start-up and as frequently as desired in operation	At prescribed intervals and when exception conditions are encountered in operation
Type of test	Mostly by attribute (pass-fail)	Mostly by variables (values)
Data retention	None or only on exception	Routine
Early recognition of failure	Yes	No
Recognition of degradation	Not usually	Yes

- The service performed is of high value;
- The value of the service depends on a future event (retrieving a file).

Built-In Test

The test capabilities that can be brought to bear on these needs at the function or assembly level are discussed in the following. Software monitoring, which is concerned with the overall functioning of the equipment, is discussed in Chapter 7. Some functions that are relatively easy to test are listed in Table 5.7.

Although Table 5.6 states that a built-in-test usually employs pass/fail criteria, these can be structured to test for quantitative performance. Thus, in the first two entries in Table 5.7, the test establishes not only that a voltage is present but also that the voltage exceeds the Zener diode threshold that can be selected to be at the minimum specified output of the power supply. In the third entry, checking the output of an oscillator with an R-C reference, an accuracy of about 5% can usually be obtained with inexpensive components. Where higher accuracy is required an independent time base in the BIT equipment may be employed. Commercial products are available to simplify BIT implementation in complex equipment [9].

Functions that are difficult to evaluate with BIT are those associated with sequencing in normal operation and particularly those associated with exception handling. Where equipment is restarted frequently, these tests can be incorporated in the start-up BIT sequence. Where this is not possible, they can be part of the testing on the field test set.

Table 5.7
Examples of Easily Testable Functions

Function	Test Mechanization
DC power supply	Zener diode of minimum spec value, followed by amplifier
AC power supply	Back-to-back Zener diodes, followed by amplifier and rectifier
Oscillator, time base	Counter with resistor-capacitor recharge circuit
Filter	Amplified ripple and noise
Counter, arithmetic unit	Inject test sequence at start-up
Status register, display	Cycle at start-up
Motor, turbine	Tachometer and detector
Bearing	Temperature sensor and/or microphone
Fan	Air vane
Heater	Temperature sensor

Field Test Set

The FTS is employed where it is important to obtain an estimate of future equipment capability. The greatest need for this arises where:

- Access for service is difficult;
- Operation depends on expendable or degrading materials (batteries, catalysts);
- Components are subject to wear out or fatigue (bearings, leaf springs).

For this type of test, the equipment must be removed from service. This permits much more flexibility in the test activities than is possible with BIT. In particular, stimuli can be inserted to cause entry into selected operating modes and to simulate failure conditions. The time for a motor to reach a specified speed can be determined, a sensitive indicator of some causes of failure.

The recording of quantitative test data (actual values of output voltages, frequencies, or mechanical forces) in the FTS permits construction of time histories that can indicate impending failures. The analysis of time trends can be handled within the FTS or it can be assigned to a central monitoring station for the assessment of an entire equipment population.

5.6 Chapter Summary

In the beginning of this chapter, we explained the difference between reliability demonstration tests and testing for reliability margins for defined performance parameters. The reliability demonstration is typically run on a pass/fail basis (testing by attributes), whereas reliability margin testing looks at the actual values (testing by variables). The latter technique can achieve a meaningful reliability assessment with fewer test hours and is therefore frequently preferred. But it addresses only the parameters for which measurements are taken and it usually does not cover failures due to lead breakage or other mechanical defects.

In later sections, we looked at tests that are part of the normal development cycle and product deployment. Because the high cost of tests makes it difficult to run specific tests for reliability assessment, it is necessary to extract as much reliability information as possible from other test activities. It is seen that there are indeed many such opportunities. An indispensable tool for using test results for reliability analysis is a FRACAS and an FRB that collects and analyzes the data and takes action to prevent reliability problems in service.

The last section described reliability data that can be obtained from testing while the equipment is in the field. BITs and the use of field test sets were described. For the latter, and also for EDCs, trend analysis can play a major role in identifying precursors to failure.

References

[1] Lloyd, D. K., and Lipow, M., "Reliability: Management, Methods, and Mathematics," Second edition, Redondo Beach, CA: published by the authors, 1977.

[2] "RADC Reliability Engineer's Toolkit," Systems Reliability and Engineering Division, USAF Rome Air Development Center, 1988. Now available from Reliability Analysis Center, Rome, NY.

[3] Freudenthal, A. M., "Safety of Structures," *Proceedings of the American Soc Civ Engrs*, Vol. 71, No. 8, October 1945, pp. 1157–1191.

[4] Hecht, H. and M. Hecht, *Reliability Prediction for Spacecraft*, RADC-TR-85-229, USAF Rome Air Development Center, December 1985.

[5] Chan, C. K., C. I. Saraidaridis, and M. Tortorella, "Sequential Sampling Plans for Early-Life Reliability Assessment," *Proc. Ann. Reliability & Maintainability Symposium 1997*, pp. 131–135.

[6] Heimann, D. I., and W. D. Clark, "Process-Related Reliability-Growth Modeling—How and Why," *Proc. Ann. Reliability & Maintainability Symposium 1992*, pp. 316–321.

[7] Hamming, Richard W., *Coding and Information Theory*, Englewood Cliffs, NJ: Prentice-Hall, 1980.

[8] Roman, S., *Introduction to Coding and Information Theory,* New York: Springer Verlag, 1997.

[9] VME Microsystems International Corporation, "Analog I/O Products (Built-In Test) Configuration Guide," Huntsville, AL, 1995.

6

Redundancy Techniques

At the beginning of Chapter 5, we remarked that it is expensive to achieve reliability by testing. The cost of test activities is primarily incurred during development. In contrast, achieving reliability by redundancy is not as costly during development but is usually very costly in production (every unit produced incurs the cost) and in operation and maintenance. The other significant difference in reliability improvement between testing and redundancy is that testing protects primarily against systematic failures (inherent in design or process), whereas redundancy protects primarily against random failures. In that respect, the two techniques are complementary.

Where random failures predominate, such as in most electronic equipment, redundancy is usually the only way in which low probability of failure can be achieved and demonstrated. A disadvantage of redundancy, in addition to the added cost, weight, and power, is the higher overall failure rate (increased maintenance expense) because more equipment is exposed to failure.

This chapter discusses material that is applicable to all or most implementations of redundancy. Broad classifications of redundancy are discussed, including detailed evaluation of several of the most frequently used structures, as well as alternatives to physical redundancy. A summary concludes the chapter.

6.1 Introduction to Redundancy at the Component Level

In Section 2.3 we introduced block diagrams of components being operated in parallel for the purpose of maintaining service when one of the components failed. This section explores this concept in greater depth. For the time being we assume that there is a perfect voting or selection mechanism and that all replicates are constantly powered, fail at the same rate, and are not repaired. Reliability trends for these conditions are shown in Figure 6.1 for dual and triple redundancy; the graph for the nonredundant (simplex) system is also shown in the figure. The failure rates for the three configurations were adjusted so that they all have the same reliability at 0.6 time units. The simplex graph represents the exponential failure law (2.2). It is clearly seen that the graphs for the two redundant configurations deviate from the exponential. Thus, a quantitative comparison between various degrees of redundancy must be referenced to a specific time interval. As shown in the figure, a high degree of redundancy is particularly effective for operation over short periods of time.

In Figure 6.1 the failure rates were adjusted to provide a cross-over in the middle of the figure so that the differences in the shape of the curves became evident. In Figure 6.2 the failure rates are identical for all three configurations so that the reliability benefits become more evident. Failure probability is, in most cases, a better criterion than reliability for the evaluation of alternatives. Thus, when we say A is twice as reliable as B, we really mean that the failure probability of A is one-half that of B.[1] Therefore the presentation of the same data in Figure 6.3 is even more meaningful.

At 0.2 units of time the triple redundant component has a failure probability of 0.01, the dual redundant one of 0.03, and the simplex one of 0.2. A

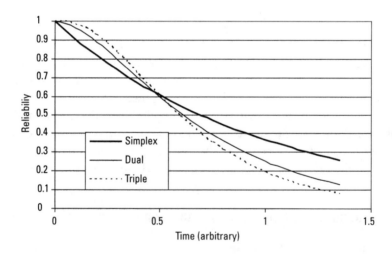

Figure 6.1 Reliability trends for redundant configurations (failure rates adjusted for equal reliability in mid-range).

---

1. A basis for this assessment is presented in Chapter 9.

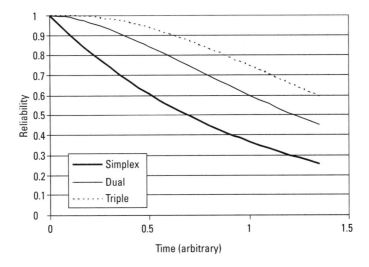

Figure 6.2 Reliability trends for redundant configurations (identical failure rates).

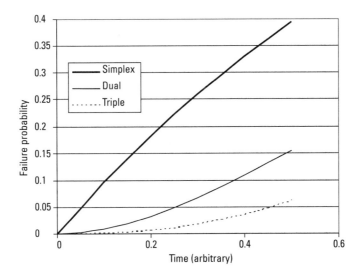

Figure 6.3 Failure probability from Figure 6.2.

failure probability of 0.2 is usually not acceptable in a design for a critical application. Thus, in this example at least dual redundancy will be required.

Another topic that affects all forms of redundancy is to select a partition level, as exemplified in Figure 6.4. We can think of component A as an array of four disk drives or a memory composed of four identical blocks. In Figure 6.4(a) the entire array is duplicated and the redundancy provisions are applied to the

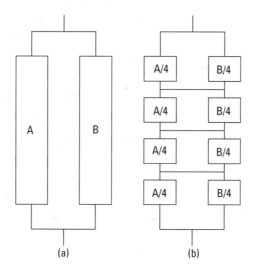

Figure 6.4 Whole and partitioned component redundancy: (a) component redundancy, and (b) partitioned redundancy.

entire component. In Figure 6.4(b) each disk drive or memory block is made redundant. In Figure 6.4(a) the entire function will fail if any disk drive in A fails at the same time as a disk drive in B. In Figure 6.4(b) the function will survive as long as at least one disk drive in each partition remains operational; it is therefore more reliable. This increase in reliability is paid for by the increased complexity of the redundancy provisions (such as switching). A comparison of the failure probability of the two configurations is shown in Figure 6.5.

For the idealized conditions (perfect switching), the failure rate of the partitioned configuration at 0.4 time units is only one-quarter that of the whole configuration. For equal partitions this advantage is easily demonstrated:

For redundancy applied to the whole the system failure probability

$$F_W = f^2 \tag{6.1}$$

And for the partitioned configuration it is

$$F_P = 4 \times (f/4)^2 = 4 \times f^2/16 = F_W/4 \tag{6.2}$$

Partitioning is not restricted to components that have identical elements. For example, in computers designed for long-duration space missions, the processor, memory, and input/output (I/O) units can be made individually redundant in order to take advantage of the higher reliability that can be achieved by

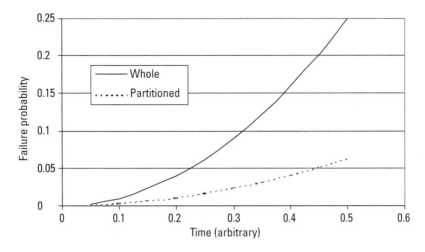

Figure 6.5 Effect of partitioning on failure probability.

partitioning [1]. A similar partitioning strategy is employed in the Stratus computer architecture to achieve extremely high availability (see Section 6.4).

6.2 Dual Redundancy

Redundancy can be applied in many ways, and the following classifications discussed in this section provide a convenient way of studying these different ways:

- Static and dynamic redundancy;
- Identical and diverse alternates;
- Powered and dormant alternates;
- Maintained (with repair) and nonmaintained applications.

6.2.1 Static and Dynamic Redundancy

In static redundancy no switching is required in order to make an alternate element available; in dynamic redundancy a switching operation (or equivalent) is required. Figure 6.6 shows a simple example of each. The objective of both structures is to provide a source of dc that is protected from failure in the power supply circuits (P. S. A and P. S. B). In static redundancy this is accomplished by using isolating diodes, and in dynamic redundancy a selector switch accomplishes it. The use of a switch (or, in more general terms, of a decision-making element) is the characteristic feature of dynamic redundancy.

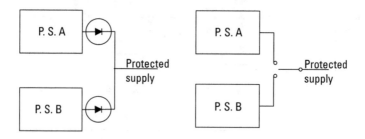

Figure 6.6 Static and dynamic redundancy.

The dc power supply used in this example is a particularly suitable candidate for static redundancy because isolation can be provided by diodes. However, the diodes do not protect against higher-than-allowable output from either of the supplies. If that needs to be prevented, additional protection has to be supplied. For mechanical components, the floating link shown in Figure 6.7 offers a convenient way of implementing static redundancy. Static redundancy by means of voting is discussed in Section 6.3 in connection with triple modular redundancy (TMR). Whenever the highest output value can be assumed to be correct (or at least preferred to the lower one), static redundancy can be applied fairly easily; similarly, where the lowest output value is always preferred. But where a signal can fail either low or high, furnish the wrong frequency, or appear at the wrong time, a decision-making element is usually required to isolate the failed component. Thus dynamic redundancy is the more flexible and widely used form.

Dynamic redundancy requires both a switching element and a fault detector for initiating switch operation (a human or automated decision maker).

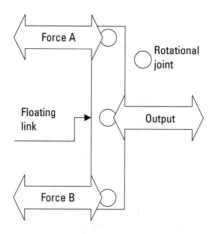

Figure 6.7 Static redundancy for mechanical elements.

These increase the cost and failure probability but usually the increments can be kept at a tolerable level. The greatest design challenge is usually posed by the error detection element.

Error Detection

The following deals exclusively with automated error detection. The reliability of humans as decision makers is too specialized to be treated in this volume [2]. Many of the self-test variables shown in Table 5.6 can be used for error detection in support of dynamic redundancy. In addition, EDCs are a powerful tool for use in digital operations, including digital communications. Attempts to access nonexistent or protected memory locations or enter an improper sequence (like a phone call going from "busy" to "ringing") are important error-detection means for digital processors.

When two powered units are present, a comparison of their output furnishes a generally applicable error detection scheme, as shown in Figure 6.8. As long as the two outputs agree, the switch can be in either the upper or lower position, but when they do not agree the switch is placed in the neutral position, thus disconnecting both units. For the floating link shown in Figure 6.7, the angle between the link and output member acts as a comparator. Disagreements between the two inputs will result in the link not being at a right angle with the output.

Where the application can accept "fail-stop" conditions (e.g., where reversion to a manual operation is possible), no steps beyond the comparison and disconnect are required. Thus, fault detection by means of comparison is a particularly suitable implementation for such an environment. Where fail-stop is not acceptable, the following additional measures may be considered:

- Put each of the units into a self-diagnostics mode (similar to a BIT on start-up, as discussed in Section 5.5) and restore a unit that passes the test to service;

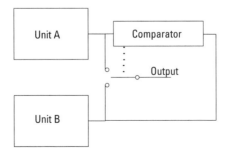

Figure 6.8 Error detection by comparison.

- If the downtime required for the previous item cannot be tolerated, switch in a spare (a unit not part of the comparison). This "pair-and-spare" scheme is discussed in Section 6.3.2.

These error detection mechanizations have evolved over the years and are constantly being improved, but they are not perfect. Estimates of their effectiveness, generally referred to as *coverage*, range from 0.95 to 0.99. In the following, we discuss how coverage can be accounted for in the evaluation of a dynamically redundant system.

Accounting for Imperfect Coverage

The state transition technique, introduced in Figure 2.5, serves as a starting point. We consider a two-unit dynamically redundant system, such as the one shown in Figure 6.8. The state designations in Figure 6.9 represent the following:

0 = both redundant units operational;
1 = one unit failed;
2 = remaining unit operational (after correct error detection);
3 = failed state (an absorbing state; no repair).

The transitions, indicated in parentheses along the arcs, are the following:

(0,0) = both units remaining in operational state;
(0,1) = failure of one unit;
(1,2) = correct error detection;
(1,3) = incorrect error detection, leading to system failure;
(2,2) = second unit remains operational;
(2,3) = second unit fails, leading to system failure.

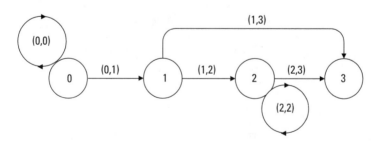

Figure 6.9 Accounting for coverage.

The transition (0,1) occurs at a rate of 2λ (either one of two units fails) and the transition (2,3) occurs at a rate of λ. The transition (1,2) is the coverage parameter and is assigned values of 1, 0.99, and 0.95 in the following.

The effect of imperfect coverage on failure probability is shown in Figure 6.10 for an assumed failure rate of $\lambda = 10 \times 10^{-6}$ per hour for a single element. Cov = 1.0 corresponds to perfect operation of the error detection and switching provisions. Less than perfect overage increases the failure probability, but for the parameter range shown here the effect will be tolerable in most applications. For the $\lambda = 10 \times 10^{-6}$/hr assumption the failure probability for a non-redundant element is 0.05 (the upper limit of the ordinate in the figure) at $t = 5{,}000$ hours. Thus, redundancy can provide a huge gain in reliability with practically achievable coverage.

6.2.2 Identical Versus Diverse Alternates

As previously stated, redundancy is particularly effective where random failures predominate. All preceding examples were based on purely random failures, and in these cases it does not matter whether the alternate elements of the redundancy structure are identical or diverse. For the sake of simplicity we have assumed identical failure probabilities for the alternates without specifying that they are also physically identical. In practice it must be recognized that some of the random failures may be due to common causes (e.g., design or manufacturing flaws) that may affect all elements (alternates) of a redundant structure. It is not necessary for failures in redundant elements to occur at exactly the same time in order to be classified as due to a common cause; the common-cause

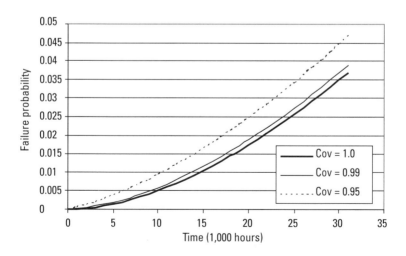

Figure 6.10 Effect of imperfect coverage.

classification means that failures due to a given cause in both elements are more likely than those calculated for random causes over the same time interval.

Use of diverse alternates is one way to minimize the effect of failures due to design but it incurs a considerable cost: separate specifications, qualification tests, test fixtures, maintenance provisions, additional spares, and training of operating and maintenance personnel. Thus it is employed only where circumstances are compelling. Spacecraft are one area where redundancy using diverse elements is frequently considered appropriate. One of the reasons is that the space environment is still not completely known and therefore the test conditions for establishing adequacy of a given design may be imperfectly specified.

The effect of nonrandom failures on redundancy is compared to that of fully random failures in Figure 6.11. The curves are constructed from the relation

$$F = 2 \times f \times (1 - Rand) + Rand \times f^2 \qquad (6.3)$$

where F is the failure probability of the redundant structure and f is the failure probability of one of its elements. *Rand* is the fraction of failures that are truly random (not due to a common cause). For *Rand* = 1 the equation is equivalent to the second part of (2.4).

The deviation from the perfect redundancy curve (heavy line) is somewhat greater than in the case of imperfect switching because common-cause failures affect both redundant elements. But the reduction in failure probability due to redundancy remains substantial, even at the high end of our common-cause failure assumption.

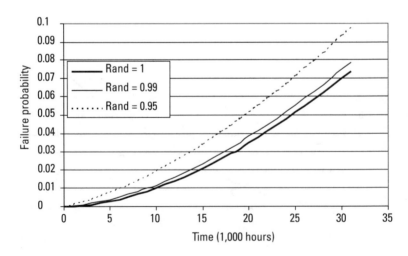

Figure 6.11 Effect of nonrandom failures.

6.2.3 Active Versus Dormant Alternates

In static redundancy all elements must be powered; in dynamic redundancy this is not necessarily the case, although it has been assumed in the calculations in this chapter up to now (by assigning the same failure rate to all elements of a redundant structure). That dormancy (being in the unpowered state) reduces the failure probability is only one of its advantages. Others are reduction in power consumption and cooling provisions. Both of these are very significant in spacecraft applications. Another motivation for keeping alternates unpowered is wear-out or depletion (e.g., bearings in mechanical devices, batteries in electrical ones).

These benefits come at a price: a longer switching interval and more uncertainty about the health of the inactive element. The switching interval is a serious obstacle in control systems for high-performance aircraft and for the powered flight of missiles and launch vehicles. It can usually be tolerated in control systems that can be brought to a safe state. The uncertainty about the health of the inactive element can be mitigated by periodic reversal of the active and standby roles. When one of the elements is not powered, failure detection by comparison (Figure 6.8) cannot be used.

In the following we discuss the reliability benefits of dormant standby elements in a dual redundant structure, as shown in Figure 6.12. An example of such an application is the monitoring system for airport directional navigation aids (localizer). The output of the monitor is displayed at the control tower of the airport, where there is also a facility for switching in a standby monitor. The monitors are sometimes placed on mountaintops or other inaccessible places, so

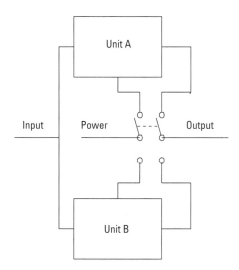

Figure 6.12 Redundancy with power switching.

that there is a substantial incentive for minimizing the number of required repairs.

The reduction in failure probability due to dormancy is represented by the dormancy factor, usually denoted d. A dormancy factor of zero corresponds to dormant units not failing at all, whereas $d = 1$ means that dormant units fail at the same rate as powered ones. For a specific electronic or electrical part, an estimate of the dormant failure rate (based on field data) can be found in the *Nonoperating Reliability Databook* [3]. An average figure of 0.08 has been suggested for the dormancy factor [4]. For our example the expected number of repairs is

$$N = (1+d) \times \lambda t \tag{6.4}$$

where λ is the failure rate for each unit, and t is the operating time for the interval of interest. Since units are repaired when they fail, the population subject to failure is constant and hence λt represents the expected number of failures per unit. Table 6.1 shows the expected number of required repairs per 10,000 hours under three assumptions of failure rates and dormancy factors of 0, 0.1, and 1 (the latter corresponding to powered operation).

While the relative (percentage) reduction in the number of repairs is independent of the failure rate, the benefit of dormancy is derived from the number of repairs that are avoided and is thus much greater at high unit failure rates. It is also seen that the savings due to dormancy are, within reason, almost independent of the dormancy factor. In this example the decision to employ dormancy would hardly have been affected even if the dormancy factor were as high as 0.2.

6.3 Triple Redundancy

Figures 6.1 through 6.3 show that going from dual to triple redundancy increases reliability. Of course, this benefit must be paid for by increased cost, weight, and power consumption of the equipment, as well as by increased test

Table 6.1
Effect of Dormancy Factor on Repair Frequency

Dormancy Factor	$\lambda = 100 \times 10^{-6}$	$\lambda = 10 \times 10^{-6}$	$\lambda = 10^{-6}$
0.0	1.0	0.10	0.010
0.1	1.1	0.11	0.011
1.0	2.0	0.20	0.020

and maintenance expense (the added third unit causes a reduction in system failure probability but at the expense of increased failure probability of units). The following discusses specific implementations of triple redundancy.

6.3.1 TMR

A widely known form of triple redundancy is the TMR with voting, shown in Figure 6.13. The input is applied to three independent elements, the outputs of which are voted. If there is any disagreement, the signal representing a majority of the units is offered as the system output. This configuration offers the following:

- Static redundancy (no error detection and switching necessary);
- Near-perfect fault coverage;
- Uninterrupted operation;
- Capability for identifying the faulty unit, thus facilitating repair.

The concept is not restricted to three modular units and the generalized form is referred to as N-modular redundancy (NMR). In an environment in which repair is not possible, the TMR structure permits continued operation as long as not more than one of the constituent units fail. System-level failure occurs when either two or all three of the units fail, thus

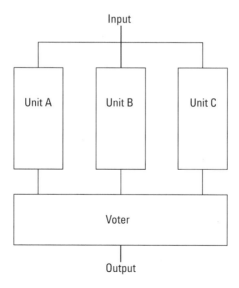

Figure 6.13 TMR with voting.

$$F = 3 \times f^2 s + f^3 \tag{6.5}$$

where s denotes unit success and f denotes unit failure. Figure 6.14 shows the time history for this equation, using the exponential reliability relation ($s = e^{-\lambda t}$ and $f = 1 - s$). The simplex failure probability is shown for comparison. The TMR structure has a much lower failure rate for the initial interval ($\lambda t < 0.1$) but this advantage decreases for longer intervals (the two curves cross at about $\lambda t = 0.7$, beyond the limit of this graph).

Repair of failed units keeps the system operating in the lower range of λt, thus achieving very low failure probabilities. Even where the equipment cannot be accessed, the benefits of repair can be achieved by remotely switching in a spare to replace a failed unit. This is called hybrid redundancy and has been considered for some space applications [5].

The TMR concept is applicable to both analog and digital systems. In analog systems a threshold for agreement needs to be defined, since the units cannot be expected to have exactly the same output values. Voting is usually implemented by median selection (i.e., the system output value is the middle one of the three values). When only two units remain, the average can be used. The vast majority of TMR applications have been in the digital field where exact agreement among the unit outputs can usually be achieved, and the output of any unit can then be used as the system output. The voter for digital TMR can be constructed from AND and OR gates, as shown in Figure 6.15. This is a single-bit voter that can be used on multibit outputs by serial transmission (sending one bit at a time).

Alternative implementations for multibit voting include loading each unit output into a register and then subtracting the registers pair-wise. Any

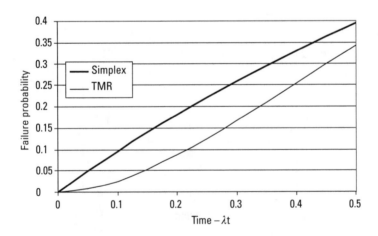

Figure 6.14 Failure probability of TMR and simplex systems.

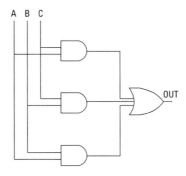

Figure 6.15 Single-bit voter.

subtraction that yields zero is a valid system output. Since the voters use fewer parts (by many orders of magnitude) than the units that are being voted, they can be assumed to be much more reliable. Where such reasoning is not acceptable, fault-tolerant voters can be used. One possible implementation is dual rail logic in which the unit outputs and their digital complements are being submitted to separate voters, one representing the negative logic of the other [6].

A common problem of all voter implementations is that they impose delay in digital processing. Part of this is due to the need to transmit data from the high-speed internal buses of the computer to an external entity over slower communication paths. The quantity of data that needs to be transmitted can be reduced by using checksums rather than the full representation of results. Another means of reducing the delay penalty is to vote only on critical outputs.

6.3.2 Pair-and-Spare Redundancy

An alternative implementation of triple redundancy is the pair-and-spare configuration shown in Figure 6.16. The outputs of units A and B are fed to a comparator, and if they agree, the output of one of them (here, unit B) is used as the system output. If the unit outputs do not agree, the comparator switches the system output to unit C. Variations of this configuration include postswitching comparisons of C with A and B and reconstituting a pair of active units. Another variation is the periodic rotation of units, so that the pair consists of A and B, B and C, and A and C in sequence.

Since the comparison is simpler than voting (which, in most cases, involves three output comparisons and then a comparison of the comparators), the pair-and-spare imposes less processing delay than TMR. Another benefit is that the spare does not need to be powered, but this is practicable only where a longer recovery time from failure can be tolerated.

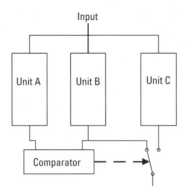

Figure 6.16 Pair-and-spare redundancy.

6.4 Higher-Order Redundant Configurations

Pair-and-Pair

An obvious extension of the pair-and-spare configuration is to switch to another pair when the comparison fails. Thus, another self-checking function is immediately available upon failure of the first pair, and the output of the new pair can be compared to that of the units of the first pair to identify the faulty one. This is the basis of the Stratus high-availability computing system that employs the concept at several layers of its partitioned architecture [7]. The failed pair can be diagnosed off-line and the good unit made available as a single spare while the failed one is being repaired.

Quadruple Voting Configurations

The TMR concept can be similarly extended to a quadruple configuration. While voting among four outputs can theoretically lead to a deadlock, the probability of this happening is extremely remote and can be accepted even in very critical applications. The flight control system computer for the U.S. Air Force F-16 fighter and the general purpose computer (GPC) for the NASA Space Shuttle are two prominent examples. In the space shuttle, there is a fifth computer that runs the Backup Flight Control software, an independently developed computer program [8]. This is a precaution to guard against problems in the common software that runs in the other four computers. A significant benefit of the quadruple configurations is that voting can be maintained even after failure of one unit.

k-out-of-n Redundancy

That four redundant units are employed when a voting configuration really requires only three is a specific instance of the *k-out-of-n* redundancy technique.

The most common example of this is that our cars carry a fifth tire that can substitute for any of the four tires that are needed. Sometimes this is referred to as a "floating spare," in contrast to the "dedicated spare" that has been the subject of most of the preceding sections.

The probability of success, P, for a system consisting of n units of which k are required is

$$P = \sum_{i=k}^{n} \frac{n!}{(n-i)!i!} \times p^{i}(1-p)^{n-i} \tag{6.6}$$

Where the exponential failure law is applicable, the substitution $p = e^{-\lambda t}$ can be made. For the frequently encountered condition where two units are required to operate ($k = 2$) and three units are provided ($n = 3$), (6.6) becomes

$$P = 3 \times p^{2}(1-p) + p^{3} \tag{6.6a}$$

The system will remain operational if at least two units remain operational and one fails (an event that can take three forms), or all three units remain operational. The failure probabilities of the 3-out-of-2 configuration will now be compared to the simplex and conventional redundant configurations shown in Figure 6.17. In Figures 6.17(a, b) the requirement that two units remain operational is shown as a series connection in the reliability block diagram. The 2-out-of-3 configuration in Figure 16.7(c) is shown as blocks connected to a selector (the circle). This is one of the frequently used symbols denoting *k-out-of-n* redundancy.

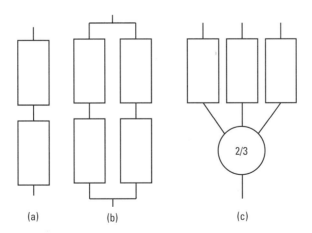

Figure 6.17 Alternative configurations: (a) simplex, (b) redundant, and (c) 2-out-of-3.

The corresponding system failure probabilities are shown in Figure 6.18. The calculations for this figure assume perfect switching and selection. Both the conventional and the 2-out-of-3 architectures have much lower failure probabilities than the simplex configuration. That the 2-out-of-3 architecture has an even lower failure probability than conventional redundancy appears surprising at first. It is explained by the greater number of possibilities for failure among the four units in the conventional redundancy.

The capabilities of *k-out-of-n* redundancy make it an important tool for failure prevention. Sometimes components are deliberately subdivided in order to permit *k-out-of-n* redundancy to be applied. An example is a high-power dc-to-dc converter for a spacecraft. In conventional redundancy, its 5-kg weight will be doubled. The converter can be redesigned for one-half of its nominal output with a 20% weight penalty so that the two halves required for operation will weigh 6 kg. But then a third half-size one is added for redundancy and the overall package will weigh only 9 kg rather than the 10 kg for conventional redundancy. A further advantage of the 2-out-of-3 architecture is that after a second failure there still remains one-half of the conversion capability, thus permitting essential loads to be supplied. Other applications of 2-out-of-3 redundancy are discussed in Chapter 11.

The selection mechanism required for *k-out-of-n* architectures generally uses the same error detection and switching means discussed earlier. The dc-to-dc converter does not necessarily need a switching mechanism at all: the three converters can be paralleled with diode isolation, similar to the static redundancy for power supplies shown in Figure 6.6. Where it is undesirable to have all three units powered, a switching arrangement similar to Figure 6.12 can be used. Comparison among the active units is a very practicable means of detecting that

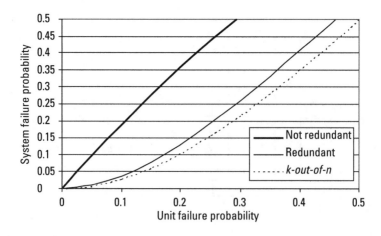

Figure 6.18 System failure probability for alternative configurations.

one of these units has failed. Replacing either one of the pair involved in the comparison with a standby unit and then repeating the comparison will identify the failed unit.

6.5 Other Forms of Redundancy

This section discusses forms of redundancy that do not require physical replication of equipment and therefore avoid the cost, weight, power, and maintenance penalties associated with the implementations covered earlier.

6.5.1 Temporal Redundancy

When we ask during a phone conversation, "Could you repeat that, please?" we are initiating redundancy in time, or *temporal redundancy*. The same process is invoked when a negative acknowledgement (NAK) is sent in an automated data exchange, or when a wrong checksum of received bits is returned to the sender of a message. These events constitute error detection, and the corrective measure is the repetition of a message. Temporal redundancy is in general very economical because unlike the measures described in preceding sections, it requires very little initial investment. The cost of retransmission is incurred only when there is an error.

Temporal redundancy is not only applicable to data transmission but to all recoverable operations for which efficient error detection can be provided. Among these are retrieval of data from storage (e.g., reading from a disk) that is protected with EDC, and arithmetic operations for which EDCs can be generated. Some common errors in sort operations can be prevented by comparing the number of sorted items to the number of input items and repeating the sort if they do not agree.

6.5.2 Analytical Redundancy

Our society generates much redundant information, and our engineering artifacts are not exempt from this. *Analytical redundancy* exploits the replication of information by substituting a source of related data for a failed source of required data. An example is that flight data systems for aircraft usually contain both airspeed and groundspeed information. Airspeed is generated from external sensors and is used by pilots or the flight control system to maintain a safe and comfortable ride. Groundspeed is generated by an inertial navigation system or by the global positioning system (GPS) and is used to predict passage over waypoints or time of arrival. In the absence of wind (and subject to some known corrections) the two quantities should be the same. The major cause for a difference is wind along the direction of flight (headwind or tailwind). This relation can be expressed as

$$G = A + W \qquad (6.7)$$

where the symbols represent groundspeed, airspeed, and wind (tailwind is a positive quantity). The contribution of the wind can be calculated as long as both A and G are available. Since wind will remain constant for short periods of time, this calculated value of W can then be used to obtain an analytical value of A when the air sensor system fails, or an analytical value of G when the primary source for it fails. The time for which substitution is valid is at least long enough to permit selection of an alternate flight plan or other means of coping with the failure.

Other potential uses of analytical redundancy in aircraft systems include barometric and radar altitude and direct measurement of angle of attack compared to pitch attitude and vertical flight path angle. Radar typically calculate the speed of a target from the Doppler component, but the speed can also be obtained from the difference of position measurements. In plant control systems pressure, temperature and fluid velocity are usually related in a way that permits calculation of one of these variables from the other two.

The primary advantage of analytical redundancy is that it does not require installation of redundant physical components in order to mitigate the effects of a failure. Analytical redundancy imposes some additional workload on a computer but that is seldom considered a significant disadvantage. It also tends to be less accurate than dedicated physical redundancy, and its applicability can be restricted (as in the case of the airspeed/groundspeed example). Where these limitations are acceptable, analytical redundancy is a valuable tool for mitigating the effect of a failure.

6.6 Chapter Summary

Redundancy offers the most comprehensive protection against random component failures. Because of its cost (equipment, installation, weight and power, maintenance) it is usually employed only for critical functions and systems. The vast majority of redundant installations involve dual or triple redundancy. Each of these can be implemented as static, where no switching is required, or dynamic (switching required) redundancy. Table 6.2 shows advantages and applicability of each approach.

Quadruple redundancy is reserved for particularly challenging applications, like the space shuttle computer or the flight controls for a high-performance fighter. Temporal redundancy and analytic redundancy provide less costly alternatives to physical redundancy for applications where some delay or loss of accuracy can be tolerated.

Table 6.2
Comparison of Redundant Architectures

	Static	Dynamic
Dual	Limited to applications that tolerate diode isolation or "wired OR" for combining signals.	Universally applicable but fault coverage depends on the effectiveness of the detection mechanism. Comparison gives good coverage, particularly for fail-stop operation.
Triple	TMR permits voting and seamless transition to a dual configuration. Near-perfect fault coverage.	Usually less maintenance than TMR but transition to dual operation is less seamless.

References

[1] Hecht, H., "Fault-Tolerant Computers for Spacecraft," *Journal of Spacecraft and Rockets*, Vol. 14, No. 10, Washington, D.C.: American Institute of Aeronautics & Astronautics, 1977, pp. 579–586.

[2] Chapanis, A., *Human Factors in Systems Engineering* (Wiley Series in Systems Engineering), New York: John Wiley & Sons, June 1996.

[3] Rossi, M. J., *Nonoperating Reliability Databook,* Utica NY: Reliability Analysis Center, 1987.

[4] Shooman, M. L., "Reliability-Driven Design of Nuclear-Powered Spacecraft," *Proceedings of the 1994 Annual Reliability and Maintainability Symposium,* ISSN 0149-144X, January 1994, pp. 510–516.

[5] Mathur, F. P., and Avizienis, A., "Reliability Analysis and Architecture of a Hybrid-Redundant Digital System: Generalized Triple Modular Redundancy with Self-Repair," *American Federation of Information Processing Societies Conference Proceedings,* Vol. 36, Montvale NJ: AFIPS Press, 1970, pp. 375–383.

[6] Krishnamurthy, R. K., and L. R. Carley, "Exploring the Design Space of Mixed Swing Quadrail for Low Power Digital Circuits," *IEEE Transactions on VLSI Systems*, Vol. 5, December 1997, pp. 388–400.

[7] Siewiorek, D., "Architecture of Fault-Tolerant Computers" *Fault-Tolerant Computer System Design,* D. K. Pradhan, (ed.), Upper Saddle River, NJ: Prentice-Hall PTR, 1996.

[8] NASA, *NSTS Shuttle Reference Manual (1988),* also available on the Avionic Systems Web site, http://science.ksc.nasa.gov/shuttle/technology/sts-newsref/sts-av.html#sts-dps-gpc.

7

Software Reliability

This chapter addresses the reliability of software, frequently referred to as *embedded software*, that is essential to operate a system. If the software is part of a feedback control system it must respond quickly and is then called *real-time software*. An example of real-time software is in the flight control system of a high-performance aircraft that must respond within a few milliseconds. An example of nonreal-time embedded software is an intrusion monitoring system that requires a response within a few seconds.

The nature of a software failure is quite different from that of most hardware failures and yet, in the embedded software environment, it is necessary to develop measures of software reliability that can be factored into system reliability. This problem is explored in Section 7.1. The conduct of software tests and the interpretation of software test results also differ significantly from hardware practices. This is discussed in Section 7.2.

Next we will turn our attention to past investigations of software failures to piece together factors that need to be controlled to prevent future ones. Section 7.3 reviews practices for preventing software failures and finds that they are not necessarily aligned with the known causes. In Section 7.4 software fault tolerance techniques are surveyed. Approaches to software reliability modeling are described in Section 7.5, followed by the chapter summary.

7.1 The Nature and Statistical Measures of Software Failures

When a capacitor shorts or a relay is stuck the only possible way to restore service is to remove and replace the failed item. Unless a redundant path exists, the system will be out of service until the failed component has been replaced. In some cases, if repeated failures of the same part have occurred, a design change will be undertaken; it may become effective as a general retrofit or as a selective retrofit only for units undergoing repair. In the case of software failures, service can often be immediately corrected by restarting the program or computer, but permanent correction always requires a design change. Since the design change may not always receive high priority, software that is known to be defective by the developer may be used routinely by the operator, who is usually not aware of the defect [1].

The bright side of this picture is that once a software fault has been successfully corrected it will never again cause a failure. Thus, "test and fix" is a much more powerful means of reliability improvement for software than it is for hardware. Permanent fault removal is also the basis of software reliability growth models, which are discussed in Section 7.4.

Given that there are fundamental differences between software and hardware failures, we must still acknowledge that both can contribute to system failures and downtime. A manager of a newly installed plant environmental control system is confronted with the problem history shown in Table 7.1, and decides that there is a serious software reliability problem.

At a meeting to formulate a plan to deal with the increase in software failures, the software development manager presents the data shown in Table 7.2 and points out that the number of faults (software problems that required correction) had actually declined over the same period.

The classifications of software change requests shown in the headers of this table are fairly common. Adaptive changes are those necessary to deal with installation of an additional temperature sensor or control console. Perfective changes are those requested to improve the system's operation (e.g., to reduce power consumption). When comparing the software entries in Table 7.1 to

Table 7.1
Monthly Problem Summary

Month	Hardware	Software
January	18	11
February	14	12
March	18	14
April	15	16
May	12	20
June	11	24
July	13	23

Table 7.2
Monthly Software Change Request Summary

Month	Total	Corrective	Adaptive	Perfective
January	9	8	1	0
February	10	8	2	0
March	10	7	2	1
April	11	5	3	3
May	12	6	2	4
June	14	5	3	6
July	13	5	4	4

those in Table 7.2, neither the number nor the direction of change agrees. The primary reason for this is that a given fault (corrective change request in the software development manager's chart) can give rise to any number of problem reports, during the time it takes for the changes to be developed, tested, and installed on the computer. The growing discrepancy between problem reports and corrective change requests indicates that there is a longer delay, possibly caused by the increase in the number of other changes being requested. When a software-enabled system is introduced into service, the operators will experience difficulties or make mistakes in some operations. Whether the changes to overcome these problems should be classified as "corrective" or "perfective" can become a subject of contractual disputes; for failure prevention, the important issue is that the corrections are made as soon as possible.

This example shows that there is no one "true" measure of software reliability. To reduce the incidence of failure in this example, a dual approach may be required: the system manager takes responsibility for preventing further activations of a problem once it is identified, thus reducing the number of problems experienced due to a single software fault, and the software development manager gives highest priority to those faults that are likely to generate a high number of problem reports. To monitor the results of such an approach, the systems manager may want to append a column to Table 7.1 that lists the new number of software faults identified, and the software development manager may want to add a column that lists the number of system failures due to software.

The preceding example dealt with a single software installation in a plant where there was interaction between the source of problem reports and the organization responsible for fixing the problems. The difficulty of measuring software reliability is increased when there are several installations of a given program. The locations may differ in the length of the workday, practices in securing the computers during nonwork hours, and in the problem reporting system. Even greater hurdles are encountered when comparing the reliability of different programs (e.g., to evaluate the effect of development methodology). We will look at several approaches but none of these satisfies all measurement needs.

Table 7.1 lists the number of reported problems per calendar month, which had been the most common measurement until the mid-1980s. Quite a few years earlier, Musa had begun to question whether calendar time was really a good measure, given that software does not fail when it is not executed, and the number of executions per month could vary largely and unpredictably [2]. An execution time measure of software reliability, such as failures per 1,000 hours of execution, began to be accepted after the publication of an expanded version of Musa's investigations in book form [3]. The execution time measure rationalizes the failure experience in a given environment. When used to compare programs running on different computers, it must be adjusted for computer speed. The computer running at higher speed performs more executions per hour and thus exposes the program to more opportunities for failure.

Another factor that enters into the evaluation of software reliability data is the scope of the program that was actually executed. The environmental control system in our earlier example may never execute the heating part of the code between April and October. The overall program may accumulate 2,000 hours of execution time but the failure rate reported during that interval is irrelevant to the reliability of the heating segment. There is much less uniformity in the measures for operational software reliability than there is in hardware reliability. The *American National Standard Recommended Practice for Software Reliability* lists six different expressions for software failure rates, most of which allow for a choice of units [4]. It is therefore necessary to be very detailed and precise when specifying or comparing failure rates.

Let us now turn to the data in Table 7.2, particularly the column that lists the corrective maintenance requests. The entries in that column represent faults (i.e., code that needs to be corrected) that are a measure of software quality but also relevant to reliability. In hardware the quality of a product is often expressed as percent defectives or per-lot defectives. The equivalent software measure is called *fault density* and is usually expressed as faults per 1,000 lines of noncomment source code. But where in hardware the percent defectives represent the results of a process that is expected to continue into the immediate future, the faults found are expected to be corrected and therefore do not represent a direct projection into the future. This objection is usually deflected by the example of 10 mousetraps placed in each of two locations. In the first location, only one trap is sprung and in the other all 10 traps are sprung. While it is true that none of the 11 mice caught will ever nibble on cheese again, it can still be assumed that the location in which the 10 traps were sprung has a greater likelihood of harboring additional mice than the one in which only one trap was sprung. Similarly, we use the faults that were found as proxies for the ones expected to be still in the code. It is not a quantitatively accurate measurement of current fault content (and likelihood of failure) but it can be used for order-of-magnitude comparisons of software quality if the following precautions are observed:

- The interval for counting faults should include comparable stages of development—the results will obviously be skewed if faults in one program include those found in early reviews and the other one does not. Also, to be statistically representative, at least 10 faults should be included for each program.
- The rigor of review and test should be comparable—a small number of faults can be due to two circumstances: the software is of exceptional quality, or the process by which faults are identified is not very thorough.
- Faults identified by field failure reports should come from the same data source or at least be reported using the same format.

Conceptually, the fault density can be transformed to software failure rate by a *fault exposure ratio, K,* in the relation

$$\lambda_S = K \times d \times v \qquad (7.1)$$

where λ is the failure rate in execution time, d is the fault density, and v is the execution speed. Musa lists values of 10^{-7} to 10^{-6} for K in the telephone switching environment [3]. The fault exposure ratio is low because of the following:

- A given instruction is not accessed at every execution (as we will show later, faults are particularly likely in instructions that are rarely accessed);
- The instruction may not yield a wrong result (cause a failure) every time it is accessed (e.g., only for negative values of one argument).

There are limitations to using (7.1) to assess the failure rate of a practical program: values of K are not known for many application areas, and the fault density, d, is subject to the possible error sources previously mentioned.

We return to the problem of accounting for software reliability in an environment where the user views the system as a whole and wants to know the failure rate and availability. In many situations the models of Figure 7.1(a) will suffice. The reliability of all system hardware elements is in series with the

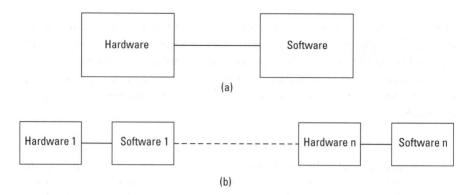

Figure 7.1 Combining hardware and software reliability models: (a) system model, and (b) subsystem model.

reliability of all system software elements, which yields the expression for system reliability

$$R = R_h \times R_s \tag{7.2}$$

This is a special case of (2.3), and where failure probability is low the approximation of (2.3a) can be used to yield

$$R \approx 1 - (F_h + F_s) \tag{7.2a}$$

where F_h and F_s are the hardware and software failure probabilities, respectively.

When software is used to monitor the operation of a hardware item the probability, R_m, that the system either operates correctly or that its failure is correctly identified is given by

$$R_m = 1 - F_h \times (1 - F_{sm}) \tag{7.2b}$$

where F_{sm} is the failure probability of the monitoring software.

7.2 Software Testing

A test report that states, "the unit under test did not fail under any of the applied test conditions" does not necessarily mean that the item is free of faults. The qualifying clause "under any of the applied test conditions" is very important. If the applied test conditions did not represent the full range of requirements, the unit may have passed the test with flying colors and still be deficient or fail in service. These remarks are applicable to both hardware and software but they are much more important for software because it:

- Operates in a digital environment, where the correct result for $x = 5$ and $x = 10$ does not ensure that the correct result will be obtained for $x = 7.5$;
- Offers much greater flexibility, such as modifying the response for various conditions, and is therefore more difficult to test for the full range of requirements.

These limitations of software test are crucial for formulating a test plan and interpreting test results. The following experience from the testing and operation of a deep space telemetry system provides an example of these difficulties. The system includes more than a million lines of code, some of which were newly developed while others were legacy or modified legacy code. It was suspected that some sections of the program were not sufficiently tested and a small-scale investigation was authorized. Programmers were asked to select segments of frequently used and rarely used code of equivalent maturity. The frequently used code represented routine address decoding, assembly of multipart text, and routing programs. The rarely used code came from initialization, calibration, and exception handling functions. The investigation's findings are shown in Table 7.3.

The fault density computed from test results (shaded row) indicates that the rarely executed segments were of higher quality (lower fault density), but in operation the rarely used segments experienced a greater number of failures, even though the frequently executed segments were exposed to greatly more

Table 7.3
Comparison of Frequently and Rarely Used Code

Characteristic	Frequent	Rare
Program size, KSLOC (1,000 source lines of code)	185.1	144.3
Number of faults found in test	893	135
Fault density from test	0.0048	0.0016
Failures during first year of operation	33	42
Failures during last 4 months of year	9	32

execution time. As might be expected, the failures in the frequently executed code occurred earlier during the operational period; rarely executed code took much longer to get debugged and probably still contained many residual faults at the end of the first year. In view of the much higher number of failures due to rarely used code, it can be concluded that the test strategy did not provide high coverage for these segments. The results also illustrate the statement at the beginning of this section that low fault density can signify either good software or incomplete test coverage.

If faults that remain after test and initial operation are concentrated in rarely executed code, it follows that failures are triggered by rare events (REs) that cause these segments to be executed under conditions that differed from those anticipated by the developers [5–7]. One of the unanticipated conditions may be high workloads (see Insert 7.1) Our focus now shifts from code segments that are infrequently executed to specific REs that produce failures and can be studied in order to produce better test cases.

A study of Space Shuttle Avionics (SSA) software failures clearly shows that REs are the predominant triggers of serious failures in well-tested software [8]. The data that will now be examined relates to failures in final (Level 8) testing of SSA software over a period of about 18 months immediately following the Challenger accident. An event was classified as rare when it had not been experienced during the preceding 20 launches. Examples of REs are engine failure, a large navigational error, or an incorrect and harmful crew command (see also Insert 7.2). When there was any doubt whether the description related to an

---

Insert 7.1—Why Computers Fail at High Workloads

More memory is in use, including address areas that had not been accessed previously and that may give rise to checksum failures.

More frequent memory use provides more opportunity for "nearest neighbor" failures (activity in a memory cell causes an upset in an adjacent memory cell).

Increased activity on communication lines causes noise on adjacent lines. Increasing code check failures and requiring retransmission—a cumulative effect.

Most code checkers detect only single bit errors. As the activity increases, so does the probability of multiple bit errors most of which will escape the checkers.

High activity causes a temperature rise, and even a small temperature rise increases the probability of semiconductor failures.

> **Insert 7.2—Causes of Rare Events**
>
> Hardware failures: computer, voter, bus, mass memory, sensors, I/O processing;
>
> Environmental effects: high data error rates (e.g., due to lightning), excessive workloads, failures in the controlled plant;
>
> Operator actions: sequences of illegal commands that saturate the error handling, unforseen command sequences (not necessarily illegal), logical or physical removal of a required component.

RE, the classification defaulted to nonrare. When at least one RE was responsible for the failure, the corresponding failure report was classified as a rare event report (RR). Many failures due to REs were associated with exception handling and redundancy management, indicating lack of a suitable test methodology for these software routines.

The SSA program had undergone intensive testing prior to the phase reported here. NASA classifies the consequences of failure (severity) on a scale of 1 to 5, where 1 represents safety critical, 2 represents mission-critical failures, and higher numbers indicate successively less mission impact. During most of this period, test failures in the first two categories were analyzed and corrected even when the events leading to the failure were outside the contractual requirements (particularly more severe environments or equipment failures than the software was intended to handle); these categories were designated as 1N and 2N, respectively. Results of the analysis are shown in Table 7.4.

REs were the exclusive cause of identifiable failures in Category 1 (safety critical). The sparse description in one report that was not classified as an RR precluded a classification by our strict criteria. In the other critical categories (1N–2N), REs were clearly the leading causes of failure but not the only one. In all critical categories, where failures were due to REs, the cause was overwhelmingly *multiple* REs (see last column, RE/RR). For the combination of the first four severity categories, the RE/RR ratio exceeds 1.95, indicating that typically the coincidence of two REs was the cause of the failure.

At this point it is appropriate to ask, "How often have we seen test cases that involved more than one rare event at a time?" The thoroughness of final testing in the shuttle program surfaced weaknesses that probably would have been detected in most other situations only after they caused operational failures. The study also showed that failures in Categories 4 and 5 are not primarily due to REs, and that in Category 3 the importance of REs is much less than in the more critical categories. One explanation for that may be that program segments that cannot cause critical failures are tested less thoroughly and therefore

Table 7.4
Analysis of Space Shuttle Avionics Software Test Failures

Severity	Number of Reports Analyzed (RA)	Number of Rare Reports (RR)	Number of Rare Events (RE)	Ratios		
				RR/RA	RE/RA	RE/RR
1	29	28	49	0.97	1.69	1.75
1N	41	33	71	0.80	1.83	2.15
2	19	12	23	0.63	1.32	1.92
2N	14	11	21	0.79	1.57	1.91
3	100	59	100	0.59	1.37	1.69
4	136	63	92	0.46	0.88	1.46
5	62	25	42	0.40	0.63	1.68
All	385	231	398	0.60	1.23	1.72

arrive at the final test still containing some of the faults that cause failures under nonrare conditions.

One further indication of the failure potential of multiple rare conditions comes from a research program sponsored by the NASA Langley Research Center to investigate the benefits of N-version programming [9]. Twenty versions of a redundancy management program, written in Pascal, were developed at four universities (five versions at each) from the same requirements, and the versions were then tested individually to establish the probability of correlated errors that would defeat the benefits of N-version fault tolerant software. The specifications for the program were carefully prepared and then independently validated to avoid introducing common causes of failure. Each programming team submitted their program only after they had tested it and were satisfied that it was correct. Then all 20 versions were subjected to an intensive third-party test program. The objective of the individual programs was to furnish an orthogonal acceleration vector from the output of a nonorthogonal array of six accelerometers after up to three arbitrary accelerometers had failed (see Chapter 11 for the motivation for using nonorthogonal arrays). Table 7.5 shows the results of the third-party test runs in each of which an accelerometer failure was simulated. The software failure statistics presented in the table were computed from Table 1 of reference [9].

Table 7.5
Tests of Redundancy Management Software

Number of Prior Anomalies	Observed Failures	Total Tests	Failure Fraction
0	1,268	134,135	0.01
1	12,921	101,151	0.13
2	83,022	143,509	0.58

The number of rare conditions (anomalies) responsible for failure was one more than the entry in the first column (because an accelerometer anomaly was simulated during the test run, and it is assumed that the software failure occurred in response to the added anomaly). In slightly more than 99% of all tests a single RE (accelerometer anomaly) could be handled, as indicated by the table's first row. Two REs produced an increase in the failure fraction by more than a factor of 10, and the majority of test cases involving three REs resulted in failure. All versions that failed under three rare conditions had already experienced failures under two rare conditions. Three-quarters of the versions that failed under two (and also under three) rare conditions had already experienced at least one failure in the runs represented by the first row. Even more encouraging: programs that had no failures under up to two rare conditions also did not fail under three rare conditions. Thus, a significant conclusion from this work is that test cases containing multiple rare conditions greatly increase the probability of finding latent faults, including those not due to the multiplicity of conditions.

A more tragic consequence of insufficient attention to testing for exception conditions resulted in serious radiation overexposures, including several fatal ones, in the operation of the Therac 25 radiation therapy device manufactured by Atomic Energy of Canada Limited (AECL) [10]. The root cause of these disastrous events probably lies in organizational failure to understand the limitations of software monitoring in safety critical situations, but in the context of this section we focus on one of the software pathways that would permit overexposure to occur. The radiation technician was required to type in the desired mode of operation and the dosage, then perform some other operations and verify the treatment parameters. If the technician discovered an error a correction could be made that was immediately accepted on the screen but that took some seconds to take effect in the machine (magnets had to be repositioned). If the subsequent steps for data entry were completed quickly, the machine would revert to the original (erroneous) settings for treatment. The handling of these exception conditions had never been evaluated in the development or the licensing procedures. Even after the accidents were being investigated by the U.S. Food and Drug Administration, its Canadian counterpart, and AECL, the combination of error correction and subsequent speedy completion of data entry was not considered creditable until a technician was asked to perform the operation exactly as on the day of the accident.

Further recognition that exception conditions are an urgent topic is seen from the following summary of the year 2000 (Y2K) software effort: "The main-line software code usually does its job. Breakdowns typically occur when the software exception code does not properly handle abnormal input or environmental conditions—or when an interface doesn't respond in the anticipated or desired manner" [11].

7.3 Failure Prevention Practices

The subject of this section represents an intersection of the disciplines of software engineering and reliability engineering. Current information on the methodology and tools that software engineering has contributed can be found in the Institute of Electrical and Electronics Engineers (IEEE) Computer Society's *Transactions on Software Engineering* [12] and in publications by the Software Engineering Institute (affiliated with Carnegie Mellon University) [13]. We focus on those methods and tools that address the handling of REs that Section 7.2 identified as a major cause of failure in systems that had been extensively screened. An important contribution to avoiding these failures could come from a better statement of requirements, and that will be this section's first topic. We will then discuss improvements of test techniques and finish with a brief description of UML-based software development that holds promise in helping with both requirements and testing. Between requirements and tests is the entire area of software design and coding, which has been omitted here because it is well covered by software engineering.

7.3.1 Requirements

Many of the failures previously discussed are frequently due to missing, vague, or incorrect requirements, particularly with regard to exception handling. System engineers and analysts are much more motivated to specify in great detail what the system should do under its normal conditions of service than what it should do (or not do) under rare conditions.

For example, consider the temperature monitoring of a high-pressure steam line. Three transducers are mounted on the line, and the software algorithm compares the readings. The following rules for furnishing a valid temperature reading are provided:

1. When all three sensors agree within a threshold, the average of the readings is used in the monitoring program.
2. When there is no agreement, the average of the two sensors that have the closest readings is used, and the other sensor is temporarily disregarded.
3. After a sensor has been disregarded in three consecutive cycles it is marked defective and it is no longer read.
4. When the two remaining sensors do not agree within the specified threshold the process is shut down for sensor replacement.

The original requirements stopped at this point. In subsequent program reviews it was realized that defective transducers cannot be replaced instantaneously and that therefore a second failure in the interval prior to replacement of a first failed sensor could not be ruled out. Because unscheduled shutdowns of the process to replace sensors is costly, it was decided to develop algorithms that could, in most cases, detect which one of the two remaining sensors was defective, based on comparing previous readings and pressure-temperature relationships. This reduced the need for shutdown to a small fraction of second sensor failure conditions. Are the requirements complete at this point?

Incomplete comprehension of all conditions associated with exception states is a common cause of missing requirements, and a number of techniques have been developed to overcome this problem. Condition tables have been used for many years [14] and are still popular because they do not depend on formal notation, are readily understood by system engineers and programmers, and can be adapted to many situations [15]. The treatment of the three temperature transducers in a condition table format is shown in Table 7.6. The instruments are labeled *A, B,* and *C.* The use of capital letters designates an operational instrument and lowercase letters designate a failed one. We assume

Table 7.6
Example of Condition Table

A	A	A	A	a	a	a	a
B	B	b	b	B	B	b	b
C	c	C	c	C	c	C	c
1	2	2	3	2	3	3	4

1. Use average of three sensors;
2. Use average of two conforming sensors;
3. Use conforming sensor if identified, otherwise shutdown;
4. Shutdown.

here that the algorithms for identifying failed sensors have been correctly implemented (these algorithms can also be stated in condition table format).

The actions to be taken for each of these conditions are identified by numbers in the table's last row and are explained in the accompanying notes. Because all eight possible combinations of sensor conditions are shown in the table, it is possible to say that the requirements have been completely stated (but not necessarily that they are correct).

A program has been described to reconstruct the condition table from delivered source code [16]. Comparing the reconstructed condition table with the original one (the requirements statement) can be done to verify that the code fully represents the requirements and that no undesirable actions have been introduced during design and coding.

Neither condition tables nor other requirements verification techniques provide assurance that the actions specified are correct for the defined conditions (see Insert 7.3). Analytical verification is possible for selected applications such as communication or security protocols. Outside of these, the correctness of actions needs to be established by reviews and by test.

7.3.2 Test

The two classical test strategies still very much in use today are functional (requirements-based) and structural (code-based) testing. Random testing can be considered a variant of either of these, depending on whether the random selection was among requirements or code segments. In an attempt to increase the "efficiency" of test, it has been suggested that test effort be allocated to segments in accordance to their expected operational frequency of invocation [17]. Such a strategy will result in minimal testing of exception handling and, if employed in a high-assurance environment, can lead to disaster [18].

Insert 7.3—Faulty Selection of Failure Criterion in a Condition Table

An earlier version of the requirements had stated the failure criterion as follows: "Any sensor that deviates by more than 5% from the average shall be declared failed."

Assume initial sensor readings at 0.5 of full scale. One sensor opens and reads 0, the others remaining at 0.5. The average of the sensors is (0.5 + 0.5 + 0)/3 = 0.33. Since each of the sensor readings differs from the average by more than 5%, all three are declared failed and the plant is shut down.

This was corrected to, "Any sensor that deviates by more than 5% from the *previous* average shall be declared failed."

An interesting glimpse into test practices for safety-critical software is provided in Figure 7.2, which shows branch coverage as a function of test cycles from tests of nuclear plant protection systems under three representative test strategies: acceptance test (functional), plant simulation (direct manipulation of the input data), and uniform random selection (among program variables). All three strategies achieve about 50% coverage on the first test case. The acceptance test progresses very slowly until about cycle 30, rises to a new plateau near 75% coverage that persists to cycle 200, and then rapidly rises to coverage above 95%. A scenario consistent with this pattern is that the first 30 cycles were used for routine test cases, mildly unusual conditions were then input up to cycle 200, and thereafter the program was subjected to rare conditions that accessed previously unused branches.

The plant simulation produces initially a higher coverage than the acceptance test, but then plateaus at about the 65% level and never goes higher. The explanation is that the plant simulator, though very capable as far as plant malfunctions are concerned, could not generate conditions that represent computer or data link failures. The uniform random strategy yields steadily rising coverage from the beginning, reaching 95% after about 50 cycles. This strategy produced a test case mix of routine, mildly unusual, and very unusual conditions, and some participants in the experiment concluded that these results show random testing to be a superior methodology to "systematic" test case generation [19]. An

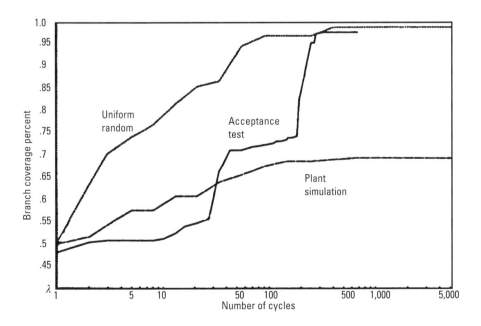

Figure 7.2 Test coverage by various strategies.

alternative interpretation is that systematic approaches need to be much more aggressive in selecting data sets that challenge the program. One approach to identifying targets for more focused testing has been described [20], but that paper recognizes that it is a concept rather than a technique suitable for general application.

An alternative approach is to use the number of REs necessary to produce a failure to assess the probability of catastrophic failures of software in operational use [21]. To understand this technique, we turn back to Table 7.4 and examine the RR/RA ratio (third column from the right). As listed in the table, this ratio decreases as we go from high severity to low severity failures. As pointed out in the table's discussion, low severity failures were observed in segments that were noncritical and therefore had received less prior testing than the critical segments. Thus, in terms of test progression, the column should be read from severity 5 (least tested) to severity 1 (most tested). In that sequence failures in early testing arise due to routine events and the contribution of REs increases until, at the end of extensive testing, practically all failures are due to REs. Further, the number of REs that contribute to the failure increases with test progression. This is evident in the RE/RR ratio shown in the last column of Table 7.4 (REs per RRs). This ratio averages 1.61 for the last three rows (early test) versus 1.93 for the top four rows (later test). Test failures that involve multiple REs are of particular interest, as will be discussed later. Rare conditions cause the program to enter code that had not previously been executed and where there is therefore a much higher probability of uncovering a fault than in previously executed segments.

To make use of this phenomenon the test cases should provide a population that is rich in individual rare conditions and the randomization among the stimuli should make it likely that multiple rare conditions will be encountered. For example, test cases may be composed of four independent stimuli that represent normal operation, N, or a rare event, R, for temperature sensor processing, radiation sensor processing, computer channel redundancy management, and software self-test, respectively. Assume that four random numbers are generated to represent the individual stimuli, and that the boundaries for N and R outcomes are selected so that for each individual stimulus there is 0.8 probability of N. The probability of encountering REs in a test case under these conditions is shown in Table 7.7. It is seen that this distribution will yield multiple RE test cases with a probability of slightly over 0.18, the sum of the last three rows.

A criterion for declaring the reliability assessment phase successful is that the *most recent X failures* that have been observed *all involve multiple REs*, and that the joint probability of encountering the multiple REs is less than the allowable failure probability of the software. The key to computing the joint probability is that the probability of the individual events, though low, will be known with much greater certainty than that of the joint event. Data for the probability of the individual events can be (and actually is) collected from a

Table 7.7
Probability of Rare Events—Four Simultaneous Stimuli, Each 0.8 Probability of Normal Operation

Number of Rare Events	Probability
0	0.4096
1	0.4096
2	0.1536
3	0.0257
4	0.0016

much larger population than that for which software failure data can be collected. There are hundreds of sensors of a given type in use in a given industry, so that failure rates as low as 0.05 per year will result in a sufficiently large number of failures to permit a good estimate of the mean time between failures.

We continue the example to show how this reasoning works for a given test case and how the X parameter can be selected. The simulated conditions that caused a specific failure are: (a) a faulty temperature sensor, and (b) failure of a computer I/O channel. In operation the temperature sensor failure is estimated to be encountered no more often than once in 20 years and computer channel failures have occurred at a rate of one in 10 years. Since these conditions occurred on different components, it is accepted that they are independent. Replacement of the temperature sensor takes 1 hour (0.0001 years), while replacement of the computer channel can occur in 15 minutes (0.00003 years). The leading causes of the joint event that produced the failure are the following:

1. The temperature sensor fails while the computer is in repair—the probability of this is:

$$P[s|rep\text{-}c] = P[s]\,P[c]\,T[c] = 0.05 \times 0.1 \times 0.000\,03 = 0.15 \times 10^{-6} \text{ per year}$$

2. The computer channel fails while the temperature sensor is in repair

$$P[c|rep\text{-}s] = P[c]\,P[s]\,T[s] = 0.1 \times 0.05 \times 0.0001 = 0.5 \times 10^{-6} \text{ per year}$$

where $P[z]$ = probability of device z failing during 1 year (z = $\underline{c}$hannel, $\underline{s}$ensor)
$T[z]$ = time to repair device z in units of years (z = c, s).

The total probability of the joint event that produced the failure is therefore 0.65×10^{-6} per year. It is thus seen that, at least for some failures caused by multiple REs, the probability of occurrence can be computed even though the probability of observing the actual failure may be negligibly small.

The cause of the particular failure just described will be corrected once it has been identified, and thus the probability of the failure is no longer of practical concern. However, if the most recent failures being experienced during a period of random testing are all due to multiple REs with a joint probability of, at most, p per year, then it can be argued that the total failure probability of the software is of the order of p, as is explained next.

Assume that the random test case generation produces one test case with multiple rare conditions for each five test cases with routine or single rare conditions (approximately the distribution shown in Table 7.7). If the probability of a failure due to a test case with multiple rare conditions is assumed to be equal to that of a test case with routine or single rare conditions, this can be represented by drawing black (single) or white (multiple) balls from an urn that contains five black balls and one white ball. The probability of drawing a white ball at the first drawing is 1/6 or 0.17, of successive white balls in two drawings (with replacement) is 0.0277, and for three successive white balls it is less than 0.005. If three successive failures due to multiple REs are observed, it can then be concluded that the probability of failure under single and multiple REs is not equal, and there is a basis for assigning a chance of less than 1 in 200 that the next failure will be due to a routine or single RE. For four successive failures due to multiple REs, the probability that this is due to random causes is less than 1 in 1,000.

This is a semiqualitative technique that may supplement other software reliability assessments or, in some applications, constitute the main test evaluation strategy. The statistical reasoning used here is not claimed to be rigorous, and the choice of a test termination criterion will involve subjective factors. But the approach offers significant practical advantages as a software test methodology where reliability requirements cannot be verified by direct observation.

7.3.3 UML-Based Software Development

In Chapter 4 we discussed the importance of FMEA in hardware reliability assessment and noted that many capabilities are lost when one attempts to use the FMEA format for software because software is not composed of discrete parts. Functional software FMEA suffers from the subjective assignment of software partitions to functions. This latter limitation can be overcome when software is specified and developed in an environment that makes use of the notation and tools of Unified Modeling Language (UML).

UML-based development tools allow us to take a fresh look at software partitioning that permits the part paradigm to be applied in generating an FMEA [22, 23]. European researchers have recognized the benefits that can be derived from UML for all forms of dependability analysis [24]. The key to this fresh look is that objects are uniquely characterized by their *methods* (sometimes called *behaviors*). At first glance *method* might be thought to be just another word for *function* but it is formally a part of the object structure and is not subject to the ambiguities we have noted for functions. As long as all methods of an object are executing in accordance with their specification, the object has not failed. Conversely, when a method does not execute in accordance with its specification, the object has failed. The failure effect will depend on the mode in which the method failed. Also, methods can frequently be partitioned in the same way that a mechanical or electronic assembly can be partitioned.

These concepts will now be explored by means of a *use case diagram* for active/standby role assignment in a plant communication system. This role assignment is a software construct that is encountered wherever components are switched from active to standby status or vice versa (dynamic redundancy, see Chapter 6). In the implementation shown in Figure 7.3 the plant control initially assigns the roles, but in steady state operation the two entities manage their roles autonomously, primarily with the aid of exchanges of heartbeat (HB) data.

The use case diagram is usually the first artifact created in UML-based development. The stick figures are called actors and are not necessarily persons. In this example the plant control may be a human or a control system. The partner is the corresponding computer program running in the partner system, and the environment represents random external events. The ovals are the specified methods and therefore the items for which failure modes are to be established. The directed lines or arcs denote information flow and hence the paths through which failure effects propagate.

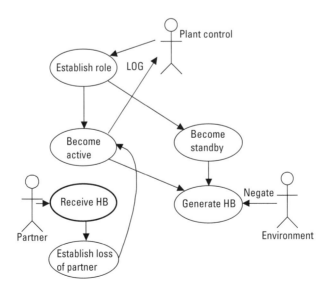

Figure 7.3 Active/standby use case diagram.

The following discussion concentrates on the receive HB method (the heavy framed oval). The most common failure mode for this method is failure to report receipt of a heartbeat when in fact it was sent by the partner. If this happens when "own program" is already in the active role, the only action is to log the failure. If it happens when own program is in the standby role, it will transition to the active role and notify plant control. Again, this is a low severity failure. But the receive HB method *may* also have a failure mode in which it signals receipt of HBs when the partner does not generate them. That failure mode may disable the entire plant communication system under the following scenario: Partner fails while in active mode. Own program does not note absence of HBs and does not take over. To determine whether this failure mode is likely to happen, we need to examine the use case diagram for the receive HB method, shown in Figure 7.4. At that level, a detailed FMEA can be generated that is comparable to the part approach for hardware FMEA.

A valid heartbeat consists of three evenly spaced pulses over a defined interval. The *counter* method transmits the number of pulses received to the *HB failure* and the *signal received* methods. If the number is three, a new interval is started by *signal received*. If it is not three, HB failure sends restart to *signal received* (permits restarting an interval) and it also stops sending alternating symbols (thus declaring failure) to the *establish loss of partner* method at the higher level (see Figure 7.3). The latter method waits three HB cycles before initiating actions appropriate to loss of partner. Thus, the failure modes that prevent

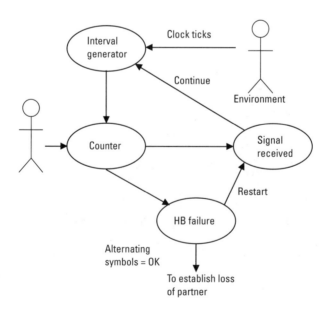

Figure 7.4 Receive HB use case diagram.

recognition of a loss of partner are (a) spurious generation of exactly three pulses per interval in *counter* and (b) spurious transmission of defined alternating symbols by *HB failure*. Both of these conditions will have to exist for at least three HB cycles to affect the actions at the *active/standby* level.

An FMEA worksheet for the receive HB method based on Figure 7.4 is shown in Table 7.8. The level of detail is compatible with a parts approach for hardware portions of the system.

The worksheet reveals both strengths and weaknesses of the receive HB method. The important strength shown is that the spurious HB failure effect will occur only under highly unlikely circumstances. An important weakness is that there is no internal fault detection. The method depends primarily on the plant control to establish whether a reported HB failure resulted from external (actual loss of partner) or internal events.

This example has demonstrated that the use of UML techniques permits generation of a software FMEA that is:

- Unique—the items analyzed are defined by the use case diagrams; the analyst is not required or permitted to partition the program into functions.
- Complete—if it is accepted that an item will work if all its methods work, then the type of analysis shown in Table 7.8 meets the completeness criterion.
- Meaningful—the failure modes examined are recognizable by system engineers and the presentation is compatible with that used for hardware.

7.4 Software Fault Tolerance

The static and dynamic redundancy architectures described in Chapter 6 can also be applied to software. One important distinction is that fault tolerance in software implies diversity because exact copies of a program are likely to fail at exactly the same instruction. Thus, a problem common to both forms of software fault tolerance is how to ensure independence of failure modes while maintaining uniformity of performance.

At some point all versions of a program will have to start at a common specification that contains the functional and timing requirements dictated by the application. And all versions will have to terminate with data structures that can be evaluated by a common evaluation mechanism. The specification or evaluation mechanism must be carefully generated and reviewed because their faults will not be covered by the fault tolerance provisions. The following have been employed to obtain independence of the software versions:

Table 7.8
FMEA Worksheet

ID	Item/ Function	Failure Mode and Causes	Local Failure Effect	Failure Detection	Compensation	Severity
1.1.1.1	Interval generator	No interval started. Loss of clock ticks or internal failure.	HB failure	External	Note 1	IV
1.1.1.2	Interval generator	Long interval. Missing clock ticks or internal failure.	HB failure	External	Note 1	IV
1.1.2.1	3-pulse counter	No count. External or internal failure.	HB failure	External	Note 1	IV
1.1.2.2	3-pulse counter	Spurious count $\neq 3$ per interval. Internal failure.	HB failure	External	Note 1	IV
1.1.2.3	3-pulse counter	Spurious count, exactly 3 per interval. Internal failure.	Spurious HB	External	Note 1	II
1.1.3.1	HB failure	Does not send restart. Internal failure.	None	External	Note 2	
1.1.3.2	HB failure	Spurious restart. internal failure.	HB failure	External	Note 1	IV
1.1.3.3	HB failure	No or random output to loss of partner. Internal failure.	HB failure	External	Note 1	IV
1.1.3.4	HB failure	Spurious defined alternating signals.	Spurious HB	External	Note 1	II
1.1.4.1	Signal received	No continue output. External or internal failure.	HB failure	External	Note 1	IV
1.1.4.2	Signal received	Spurious continue output. Simultaneous errors in input and restart processing.	None	External	3-pulse counter	None

Note 1: Temporary failure effects are suppressed because the loss of partner method waits for three intervals to activate.

Note 2: Will cause Severity IV effect under all conditions when count $\neq 3$ and no effect when count = 3.

- Independent design teams and design rules;
- Independent programming teams;
- Different computer languages and compilers;
- Different processors.

Despite these precautions, there have been problems achieving fault tolerance in some applications due to common misinterpretation of the specification, particularly as it relates to exception handling [9], and actions of the operating system that recognizes a software problem and causes it to abort before fault tolerance provisions can take over [25].

Static software fault tolerance depends on a voter that can be implemented in either hardware or software. The latter approach is more flexible in allowing small or temporary differences to be suppressed and in changing from "tight" to "loose" comparisons, depending on system conditions. But it is much slower than hardware voting and requires careful design for time-critical applications [26]. The most common form of static fault tolerance for software is referred to as n-version programming [27]. The versions can be executed simultaneously on multiple computers of the same design or of different designs. Alternatively, where time constraints permit, they can execute sequentially on the same computer or array of redundant computers.

Dynamic fault tolerance for software involves subjecting the primary version to an acceptance test and transferring to an alternate if that test fails. A robust structure for these provisions has been designated as the recovery block [28]. When the primary version fails there is inherently sequential execution with accompanying delay. Because this delay is experienced very infrequently it may be tolerable even in time-constrained applications and the amount of lost program execution can be reduced by scheduling acceptance tests at intermediate points in a program's sequence (e.g., after input data has been read or after a matrix inversion has been completed). Acceptance tests can be performed much faster than n-version voting because they are performed in the same computer as the program that they service and therefore do not have to wait for data from external computers.

Software fault tolerance is inherently expensive because it requires multiple (and preferably independent) design, development, and test, plus usually provisions for multiple computers. It is therefore reserved for extremely critical applications. The reliability assessment for fault tolerant software cannot assume completely independent failure modes for the redundant structures, and a block diagram such as the one shown in Figure 7.5 is appropriate [29]. The common block accounts for failures due to the (common) requirements and voting structure, as well as for correlated failures in the two versions. The RBD is extensible to three or more versions.

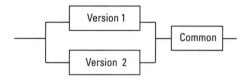

Figure 7.5 RBD for fault tolerant software.

7.5 Software Reliability Models

Section 7.1 explained that software failures can frequently be overcome by just restarting the program because the failure occurred only when a particular data value or timing relationship was encountered. When the program is not changed, the failure can be expected to recur at a frequency that is dependent on encountering the conditions that trigger the failure. When a program is in widespread use (so that local variations are smoothed out) the failure rate due to specific causes will remain constant [30].

Where faults are being removed as they are found, such as is typical in testing, the failure rate can be expected to decrease, leading to reliability growth. In most cases reference to software reliability models means the software reliability growth model. A number of these models have been defined in the previously mentioned ANSI standard [4]. Most assume that the fault exposure ratio is constant. Typical of these is the Musa-Okumoto model

$$\lambda(t) = \lambda_o \exp[-\theta t] \qquad (7.3)$$

where λ_o is the initial failure rate, θ is an empirical decay factor, and t is execution time. In contrast, the Littlewood-Verall (L-V) model allows for a changing fault exposure ratio. As discussed previously, there is frequently a transition from failures that occur under routine conditions to those that occur only under exception conditions and ultimately to those that occur under multiple exception conditions. These transitions make it likely that there are changes in the fault exposure ratio and thus make the L-V model preferable. However, the L-V model requires estimation of additional empirical parameters, and where only a small number of failures are observed, the data for this estimation is only rarely available [31].

7.6 Chapter Summary

Differences between software and hardware failure mechanisms pose problems in the assessment of digital systems. It is essential to distinguish between software faults, the correction of which can take a considerable amount of time, and software failures that can typically be corrected by restarting the program or resetting the computer. Because one fault can cause many failures it may be advisable to restrict the operation of a system until the fault has been corrected.

Once a program has successfully operated on a data set under specific timing conditions it will always process that data set correctly under the same conditions. It follows that a program that has been debugged as a result of reviews and testing is not likely to fail while processing normal data under normal conditions. Failures are more likely to happen when it processes exception conditions, particularly multiple REs.

To make a program as reliable as possible it must therefore be tested under conditions that cause a lot of exceptions to be processed. Random selection of exception conditions and a test environment that generates multiple exceptions for a given test case promise economical elimination of software faults.

The UML and software development using UML tools can eliminate omissions and ambiguities in requirements statements. It can also facilitate the generation of software FMEA with detail and objectivity that is comparable to hardware FMEA and that therefore can be used as the centerpiece of a reliability program.

Fault tolerant software can be and has been used to satisfy the highest reliability requirements in aircraft flight control and in space and missile applications. But it is an expensive technique because all versions have to be separately developed and tested. It does not eliminate failure sources in the requirements that are common to all versions and in the acceptance test or voting that follows the execution of the versions.

As faults are removed from a program, its probability of failure decreases. Software reliability models help quantify this reliability growth.

References

[1] Lee, I., and R. K. Iyer, "Diagnosing Rediscovered Software Problems Using Symptoms," *IEEE Transactions on Software Engineering*, Vol. 28, No. 2, February 2000.

[2] Musa, J. D., "A Theory of Software Reliability and Its Application," *IEEE Transactions on Software Engineering*, Vol. 1, No. 2, 1975, pp. 312–327.

[3] Musa, J. D., A. Iannino, and K. Okumoto, *Software Reliability: Measurement, Prediction, Application*, New York: McGraw Hill, 1987.

[4] American National Standards Institute (ANSI), *American National Standard Recommended Practice for Software Reliability R-013-1992*, sponsored by the American Institute for Aeronautics and Astronautics, Washington, D.C., Approved 1993.

[5] Iyer, R. K., and D. J. Rosetti, "A Statistical Load Dependency Model for CPU Errors in SLAC," *12th International Symposium on Fault Tolerant Computing*, IEEE Cat. 82CH1760-8, June 1982, pp. 363–372.

[6] Velardi, P., and R. K. Iyer, "A Study of Software Failures and Recovery in the MVS Operating System," *IEEE Transactions on Computers,* Vol. C-33, No. 7, July 1984.

[7] Kanoun, K., and T. Sabourin, "Software Dependability of a Telephone Switching System," *17th Fault Tolerant Computer Symposium,* IEEE Computer Society, Pittsburgh, PA, June 1987, pp. 236–241.

[8] Hecht, H., and P. Crane, "Rare Conditions and Their Effect on Software Failures," *Proceedings of the 1994 Reliability and Maintainability Symposium,* IEEE, Piscataway, NJ, pp. 334–337.

[9] Eckhardt, D. E., et al., "An Experimental Evaluation of Software Redundancy as a Strategy for Improving Reliability," *IEEE Trans. Software Engineering,* Vol. 17, No. 7, July 1991, pp. 692–702.

[10] Leveson, Nancy G., *Safeware,* Reading MA: Addison Wesley, 1995.

[11] Hansen, C. K., "The Status of Reliability Engineering Technology 2001," *Reliability Society Newsletter,* IEEE Reliability Society, January 2001.

[12] http://www.computer.org/tse.

[13] http://www.sei.cmu.edu.

[14] Goodenough, J. B., and S. L. Gerhart, "Toward a Theory of Test Data Selection," *IEEE Transactions on Software Engineering,* Vol. SE-1, No. 2, June 1975, pp.156–173.

[15] Parnas, D. L., G. J. K. Asmis, and J. Madey, "Assessment of Safety-Critical Software," *Proc. Ninth Annual Software Reliability Symposium,* Colorado Springs, CO, May 1991.

[16] Hecht, M., K. Tso, and S. Hochhauser, "Enhanced Condition Tables for Verification of Critical Software," *Proc. of 7th International Conference on Software Testing,* San Francisco, CA, June 1990.

[17] Liao, Shih-Sung, *An Integrated Testing Approach for Object-Oriented Programs,* Ph.D. Dissertation, Department of Computer Science and Engineering, Auburn University, March 1997.

[18] Li, N., and Y. K. Malaya, "On Input Profile Selection for Software Testing," *Proc. of ISSRE 94,* pp. 196–205.

[19] Bishop, P. G. (ed.), *Dependability of Critical Computer Systems 3—Techniques Directory,* Elsevier Applied Science, 1990.

[20] Voas, J., F. Charron, and K. Miller, "Investigating Rare-Event Failure Tolerance: Reductions in Future Uncertainty," *Proc. IEEE High-Assurance System Engineering Workshop (HASE'96),* October 1996.

[21] Hecht, H., and M. Hecht, "Qualitative Interpretation of Software Test Data," *International Workshop on Computer Aided Design,* Test and Evaluation for Dependability, Beijing, July 1996.

[22] Booch, G., J. Rumbaugh, and I. Jacobson, *The Unified Modeling Language User Guide,* Reading MA: Addison-Wesley Object Technology Series, 1998.

[23] Rosenberg, D., *Use Case Driven Object Modeling with UML,* Reading MA: Addison-Wesley, 1999.

[24] Bondavalli, A., I. Majzik, and I. Mura, "Automatic Dependability Analysis for Supporting Design Decisions in UML," *Proc. of 4th IEEE International Symposium on High-Assurance Systems Engineering (HASE'99)*, Washington, D.C., November 1999.

[25] Avizienis, A., et al., "In Search of Effective Diversity: A Six-Language Study of Fault Tolerant Flight Control Software," *Digest of Papers, FTCS-18,* IEEE Computer Society, June 1988.

[26] Wensley, J. H., et al., "SIFT: The Design and Analysis of a Fault Tolerant Computer for Aircraft Control," *Proc. of the IEEE*, Vol. 66, No. 10, October 1978, pp. 1240–1254.

[27] Avizienis, A., "The N-Version Approach to Fault-Tolerant Software," *IEEE Trans. on Software Engineering*, Vol. SE-11, No. 2, June 1981, pp. 185–222.

[28] Randell, B., P. A. Lee, and P. C. Treleaven, "Reliability Issues in Computing System Design," *Computing Surveys*, Vol. 10, No. 2, June 1978, pp. 123–165.

[29] Hecht, H., "Fault Tolerant Software," *IEEE Trans. on Reliability*, Vol. R-28, No. 3, August 1979, pp. 227–232.

[30] Adams, E. N., "Optimizing Preventive Service of Software Products," *IBM Journal of Research & Development,* January 1984, pp. 2–14.

[31] Abdel-Ghaly, A. A., P. Y. Chan, and B. Littlewood, "Evaluation of Competing Software Reliability Predictions," *IEEE Transactions on Software Engineering*, SE-12 (9), September 1986, pp. 950–967.

8

Failure Prevention in the Life Cycle

We partition the life cycle into the following phases:

- Concept;
- Development;
- O&M.

Of these, the development phase usually has the greatest impact on reliability, and it will receive special emphasis in this chapter. The target of the failure prevention effort will be referred to as the *system*. The system can be a space mission (a single-use project) or a digital control system, of which thousands of copies will be produced. In the latter case the concern is with the *product* reliability, and we refer to this as a product system. The life cycles of the single-use and product systems have much in common and will mostly be discussed together.

Reliability activities over the life cycle are usually documented in a *reliability program plan*. Of course, this plan must be integrated with the project's general life-cycle events, and therefore the first section of this chapter introduces the format and terminology of a project life cycle. After that, we describe the reliability issues that arise in the life-cycle phases. The structure and content of a reliability program plan are presented in Section 8.3. Key elements of life-cycle activities are reviews and audits, and aspects of these important to failure prevention are presented in Section 8.4. This is followed by a section on monitoring requirements. The chapter concludes with a summary.

8.1 Life-Cycle Format and Terminology

All but trivial projects involve the collaboration of several organizations and multiple individuals within each organization. To coordinate these diverse efforts a number of project management tools are commonly used [1, 2]. For our purposes, the basic concept is that of *activity*, as diagrammed in Figure 8.1. The activity starts when all prerequisites have been completed, and it ends when a review establishes that all requirements for the activity have been met. Typically, several activities are carried out at the same time, but if activity A furnishes prerequisites for activity B, they must, of course, be conducted in sequence.

A number of activities can be grouped into an *activity block* and a number of activity blocks make up a *life-cycle phase*. The end of an activity block and of a phase is usually referred to as a *milestone* that is marked by release of documents and the conduct of major program reviews. The sequence of activity blocks within a phase is frequently diagrammed in a *waterfall chart*. Figure 8.2 is an example of a waterfall model for the development phase. The review block in Figure 8.1 has been merged into the "Complete?" decision diamond. The waterfall model has been in use at least since the late 1960s [3]. It forces sponsors and users of a project or product to agree with developers that requirements, design, and so on are complete before the next activity block starts. This was seen as a good business practice since budgets and payments could be tied to a milestone. In most cases it is very difficult to have a clean completion; when the sponsor and

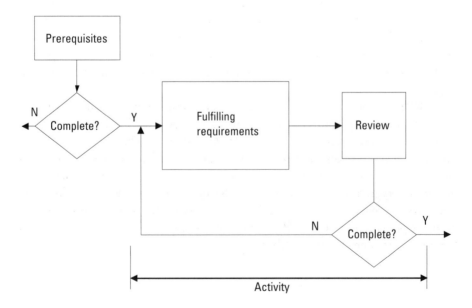

Figure 8.1 Activity template.

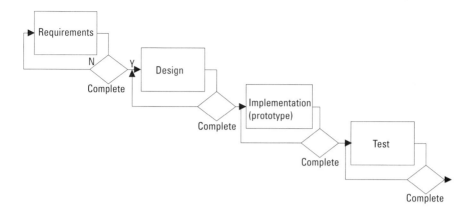

Figure 8.2 Waterfall model for the development phase.

developer cannot agree on the acceptability of an item, neither side wants to see the staff idled and the schedule slipped. Therefore, partial or conditional milestone completions are resorted to, much to the chagrin of contracts management.

Today the waterfall model is regarded as simplistic, particularly for major systems where development typically takes 3 to 5 years. It is unrealistic to expect that requirements can be frozen near the beginning of that time span, and the model does not have an easy way for phasing in revisions that can arise from changes in user needs, advances in technology, or problems encountered in later activity blocks. When a change in requirements becomes necessary, an engineering change proposal (ECP) is prepared, usually by the developer, and is reviewed and eventually approved by the sponsor. When additional funding is required (this is almost always the case), the process of ECP preparation and approval can extend over a period of many months. Informal, poorly documented agreements are frequently necessary to cope with "show stoppers"; these can lead to confusion about which configuration is currently being worked on. A NASA Web site states: "The standard waterfall model is associated with the failure or cancellation of a number of large systems. It can also be very expensive. As a result, the software development community has experimented with a number of alternative approaches" [4].

One alternative is the spiral development model originated by Barry Boehm [5] that allows for iteration of activity blocks during development. An example of the spiral development model is shown in Figure 8.3.

Like the waterfall model, the spiral model provides for review and assessment of completion at the conclusion of each activity block but these have been omitted from the illustration to keep it uncluttered. The figure shows only two iterations but three or four are not uncommon. The significant change from the waterfall model is that it recognizes the need to revisit requirements after some

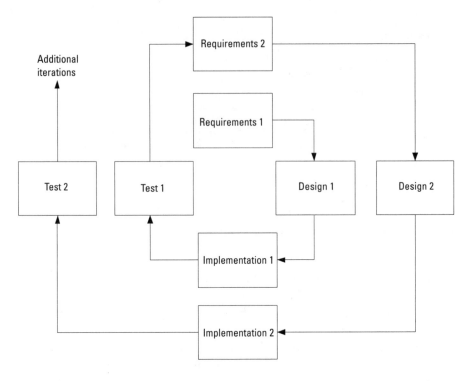

Figure 8.3 Spiral model for the development phase.

of the design, implementation, and testing have been completed and to evolve the system throughout the process. When applied by competent developers, the first iteration will include the system's most critical and innovative (risky) elements. This permits accommodation for problems encountered with these elements during subsequent iterations.

For the reliability practitioner, there is a significant difference between looking at an ECP and participating in a review of a phase in the spiral development model. The review provides the following:

- A broad analysis of the reliability aspects of a change;
- Interaction with other disciplines about the effects of a change;
- An opportunity to formalize requirements for testing or other verification measures at later development phases.

At each design phase iteration, an update of the FMEA should be scheduled. FMEA updates are practically never scheduled if the waterfall model is used, regardless of the number of changes implemented, and this causes the FMEA to eventually become irrelevant.

The spiral model is one of a number of methodologies that have come into use to counteract the premise that requirements will be complete at the beginning of the development of a major system. Others are *rapid prototyping* [6] and *extreme programming* [7], both of which are aimed at software development but can be adapted to systems heavily dependent on software. The former is preferred where the user interface represents a high-risk element and the latter where the user has little experience or motivation in requirements formulation (the user becomes part of the development team). When these methodologies are used, it is important that milestones provide for reliability reviews.

Regardless of the life-cycle model used, the documentation furnished at each milestone must be controlled so that all parties that contributed to its generation cannot change it without agreement. This process is called *configuration management* and it must straddle two conflicting aims: maintain stability in the system design and facilitate orderly change (i.e., change that is documented and coordinated with all affected parties). During the concept phase, it is usually not desirable to impose this degree of formality, particularly if participants are few in number and communicate freely. Once requirements for the development phase are released, there are many more individuals and functions involved, and the need for configuration management increases. Failure to implement and adhere to this practice has been known to cause reliability problems [8, 9].

8.2 Reliability Issues in Life-Cycle Phases

The issues are grouped by the major life-cycle phases in which they usually arise: concept, development, and O&M.

Reliability Issues in the Concept Phase

The start of the concept phase is not usually a well-marked event. There may be a *need* or an *opportunity;* frequently a combination of both. A typical need is that a user finds inefficiencies or difficulties in a current task or may want to take on a new task. An example is that most people find it unduly time consuming to reconcile their checkbook balance with the bank statement. This need led to the concept of software programs that greatly reduce the effort for this task. An opportunity may arise from a technological development that promises to simplify a currently difficult task or make possible a new line of activities. An example is the use of GPS receivers for car navigation. These ideas are usually discussed and elaborated in a small group that has no formal funding. If the concept survives scrutiny by local management, it obtains funding from a discretionary research and development (R&D) budget to prepare a formal proposal that, if accepted, will then lead to the transition to the development phase.

Only rarely are detailed reliability requirements generated during the concept phase. But at the conclusion of the phase, answers to most of the following questions pertinent to system reliability should be available:

- What life-cycle model will be used?
  - Expected start and finish of the development phase;
  - Reviews and milestones.
- Is the system hardware and software sufficiently defined so that reliability planning can start?
  - Existing components—availability of failure analyses and statistics;
  - Modified components—extent of modification and availability of reliability data;
  - New designs—extent of innovation and use of new parts and materials.
- What are the organizational interfaces?
  - Customer reliability organization;
  - Regulatory agency or standards organization;
  - Independent verification and validation (V&V) contractor;
  - Subcontractors and consultants.
- Are there special areas of safety and reliability concerns?
  - Autonomous failure detection;
  - Mechanical integrity and leak prevention;
  - Extreme temperature and vibration environments;
  - Exposure to high radiation levels;
  - Software design or reliability requirements.
- Will new design tools, materials, or processes be required?
  - Contact information for obtaining reliability relevant data.
- Have reliability or availability goals been established?
  - Customer requirements or expectations for MTBF and MTTR;
  - Product warranty policy;
  - Expected product life;
  - Anticipated maintenance and logistics organization.

Reliability Issues in the Development Phase

Starting with the development phase, the responsibilities and activities for failure prevention are usually defined in the reliability program plan discussed in Section 8.3. In this section, we list some issues that overlap the activity categories detailed in the plan. These include the following:

- System failure modes:
  - Allocation and traceability to components;
  - System-specific severity classifications;
  - Identification of precursors to critical failure modes.
- Failure histories of existing and modified components:
  - Potential design and process weaknesses;
  - Statistics of random failures.
- Reliability concerns for new designs:
  - Knowledge of failure mechanisms;
  - Evaluation of design margins;
  - Monitoring provisions for margins to critical failures.
- Failure prevention modalities:
  - Design margins for identified failure mechanisms;
  - Inspection and test as protection against process-induced failures;
  - Need for redundancy:
    - Safety and mission-critical functions;
    - To achieve high availability.
  - Opportunities for redundancy:
    - Power supplies;
    - Multiply used components (*k-out-of-n* redundancy).
  - Error detecting and correcting codes.
- Feedback mechanisms:
  - Failure reporting and corrective action systems;
  - Recording of margin monitoring provisions;
  - Recording of operation of error correcting codes.

System failure modes in the highest two severities (in many cases identified in a preliminary hazards analysis or similar document) must be allocated to components that can cause or contribute to these failure modes. Within each identified component, it may be possible to further identify functions or parts that contribute to these failure modes. This information should then be furnished to the design, quality assurance, and manufacturing organizations so that safeguards against these failure mechanisms can be provided. Within the reliability organization, the failure mode allocations, and reliability allocation in general, are compared to the component vulnerabilities obtained from the failure histories (for existing and modified components) and reliability concerns (for new components) to establish an initial reliability assessment for each component, as shown in Figure 8.4. From these assessments, a preliminary reliability estimate and failure prevention strategy can then be generated, drawing on the failure

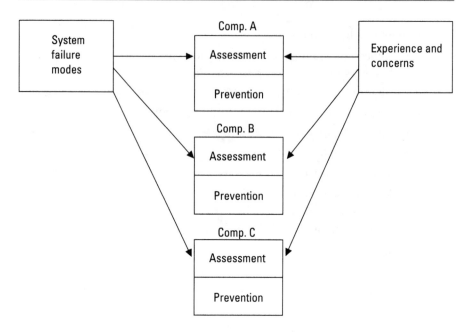

Figure 8.4 Initial reliability planning.

prevention modalities shown in the fourth bullet. This approach is consistent with emphasis on the "area of concern" discussed in Chapter 4 (see Figure 4.6). The feedback mechanisms do not directly enter into this initial planning but the need for them should be kept in mind when selecting prevention modalities.

The initial planning is usually not dependent on specific failure-rate predictions, except where use of components with known high failure rates cannot be avoided. Similarly, reliability testing and assessment need not be considered at this stage. They will be added during the development phase as the initial planning is elaborated and modified in the generation of the reliability program plan document.

Toward the end of the development phase, the emphasis of the reliability efforts shifts to a review of test results and, particularly, test failures. The activities involved are essentially the same as those that are the central issue in the O&M phase and are described next.

Reliability Issues in the O&M Phase

The central reliability activity during the O&M phase is to respond to *problem reports*. The choice of the term "problem reports" is deliberate because not every problem report pertains to a failure. Even those reports that do not relate to failures may need attention because they can indicate insufficient user training and support. There can be problem reports in which the first echelon repair facility declares that the equipment is operating satisfactorily. The terms "could not duplicate" (CND) or "field failure not verified" (FFNV) are sometimes used in such reports. Where they constitute a noticeable portion of the report population, their investigation becomes a major concern of the reliability organization. Possible causes include the following:

- Mechanical intermittencies (loose connections, solder balls);
- Electrical transients due to the power supply, adjacent equipment, or switching within the equipment;
- Software that responds in unexpected ways to some exception conditions.

In all cases, improved testability provisions by internal monitoring, remote monitoring, or external test sets is desirable.

Once a problem is found to constitute a verified failure, it may be processed, as shown in Figure 8.5. Preparing failure reports may be delegated to organizations that are not part of the developer (except in single-use projects). But the further activities now described benefit from being closely associated with the development team, usually through an FRB. The first step is to determine whether the failure is chargeable or nonchargeable. The major contributor to the latter category is usually induced failures (due to outside events or other failures within the system). Other nonchargeable failures include use of the equipment in an unauthorized manner (including physical damage), and failures in limited life parts beyond the stated life. If the failure is found to be nonchargeable, it is repaired (usually at the user's expense), tested, and returned to service after it passes testing. If the test fails, there are two possibilities:

- It is found that the repair was not carried out correctly and must be repeated.
- The repair was carried out correctly but the problem was not fixed or a new problem was created. This indicates that the diagnostic procedure was incorrect and must be revised. Repair against the revised diagnostics is then undertaken, leading to test and restoration to service.

If the failure is found not to be induced, the problem report is reviewed for consistency with the FMEA. This can be accomplished by having the report's originator list the applicable ID (from the FMEA worksheet) for this failure, or it can be assigned to the analyst as part of the problem report review. If the failure mode and effects are not found to be consistent with the FMEA, that document must be revised. At that time it should also be reviewed to determine whether the omission of this failure mode (or defective entry in one of the worksheet fields) arose from a systematic deficiency in the FMEA methodology. If so,

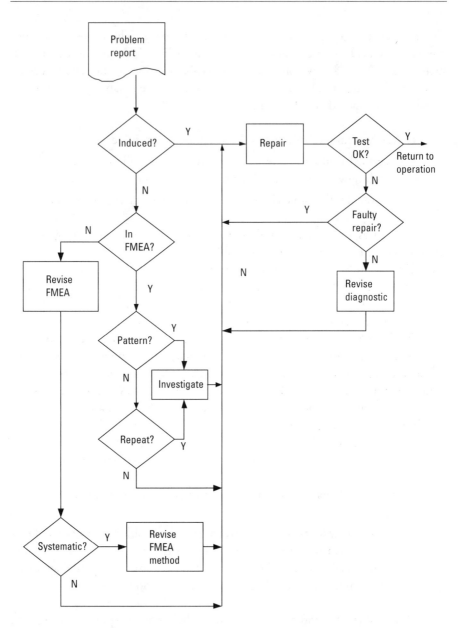

Figure 8.5 Failure processing template.

the methodology must be revised. The component can then be cleared for the repair process.

For failures consistent with the FMEA entry, two further reviews should be conducted: to determine whether there is a pattern of similar failures (in other components of the same type) or whether this component is a repeater (has been repeatedly repaired). If either of these conditions is found to be true, an investigation will determine whether a design change (for pattern failures) or a change in repair procedures (for repeaters) is required. The existence of a pattern failure can be established if there are multiple reports with a given FMEA identifier; the existence of repeat failures can be established by referring to the serial number of the failed component. The failed components then enter the repair process as indicated on Figure 8.5.

Software failures may be classified as nonchargeable if operations were outside the specified environment (e.g., heavier workload) but this is usually much more difficult to establish than for hardware failures. A confirmed software failure will always be a pattern failure (since all copies carry the same code) and will therefore require a design change. While the change is being implemented, the operation of the equipment may be restricted (placarded) to avoid entering the specific conditions that will cause the failure. The design change for software can be accomplished in two ways:

1. As a patch, is which the executable code is modified to avoid the failure condition;
2. As a revision, in which the source code and documentation are changed.

The latter is the more desirable, but also more time-consuming, procedure. In many systems, revisions are scheduled only once or twice a year. In between revisions, a higher failure rate may be accepted, the system can be placarded, or a patch can be authorized.

As indicated, software failures and pattern hardware failures may cause design changes in a fielded system. But there are several other, and usually more significant, reasons why a design needs to be changed:

- User requirements change or flaws are found in the initial requirements;
- Technological advances permit a better solution of the task than the system is designed for;
- Hardware components become obsolete and cannot be maintained;
- Interfacing systems or components are being changed.

Design revisions in response to the first two bullets are usually referred to as *perfective* changes; those responding to the last two bullets are usually called *adaptive* changes.

Each of these changes can have unfavorable effects on system reliability and availability. A frequent avenue for these challenges to reliability is that documentation and other system support (test fixtures, training, and failure reporting system) are not updated in a timely manner. The reliability staff may be under pressure to waive the immediate modification of support items, but it must be aware of the consequences that can (and usually will) ensue.

8.3 The Reliability Program Plan

A reliability program plan had been a requirement for all military systems in MIL-STD-785 [10]. The standard has been withdrawn from official use but many of the provisions can still be used as guidance for achieving reliability in a major program. MIL-STD-785 includes several provisions for reliability accounting and program surveillance, tasks intended to permit the customer to supervise the conduct of reliability activities, particularly during the development phase. We eliminated these provisions from the following discussion because they were rarely effective in failure prevention. For the development of very large systems, a reliability program with emphasis on the management component may be required, but smaller systems may be better served by a less formal structure. Table 8.1 is an adaptation of the reliability program plan from MIL-STD-785. Readers are encouraged to pick and chose items pertinent in their environment.

As indicated in the comment column, many of the program plan's elements are a matter of management style. In general, the need for these increases with the size of the project. Also, some customers may insist on having defined lines of responsibility and task control consistent with the work breakdown structure (WBS). Other elements are essential regardless of the plan's structure. Even if no reliability program plan is generated, there must be a mechanism for ensuring that these activities are carried out.

We turn our attention now to the tasks that *may* be a part of the plan, divided into three major divisions:

- Program management;
- Design and evaluation;
- Testing.

The first bullet includes preparing the reliability program plan, failure reporting, and the FRB, all of which have been discussed. It also includes program reviews, discussed in Section 8.4, and control of subcontractors and suppliers. The key elements in this latter topic are well-defined interfaces and avoiding barriers to communication. The product specification may contain explicit reliability requirements or these may be implicit (design margins, service life, and so on). Reliability concerns are how conformance to these requirements can be monitored and remedial action to be taken when there is nonconformance. The importance of communication will be recognized when requirements are not met.

Table 8.1
Elements of a Reliability Program Plan

Ref	Description	Comments
a	Reliability requirements derived from statement of work (SoW)	Needed in any plan
b	Description of each reliability task	Task structure is not always essential
c	Organization responsible for each task and completion criteria	Depends on management style
d	Interfaces between reliability tasks and with other tasks	Control of interface with other tasks is needed
e	Schedule of reliability tasks	See b and c
f	Identification of known reliability problems	See Section 8.2
g	Procedures for resolving the problems in f.	Depends on f.
h	Designation of reliability milestones	See c
i	Notification of reliability concerns to designers and others	See Section 8.2
j	Key personnel responsible for reliability tasks	See c
k	Management structure (line and staff)	See c
l	Listing of reliability guidance and review documents	See c
m	Contribution of the reliability function and its authority	See c
n	Reliability data from predecessor systems	See Section 8.2

Very few serious reliability problems are solved by legalistic interpretations of the applicable documents. Development entails dealing with unknown factors, and the subcontractor or supplier may have agreed to a specification in good faith without expecting a difficulty that has now become manifest. A cooperative effort is the best way to overcome problems that arise during development. A frequently encountered reliability problem is that the internal temperature rise in a circuit board is greater than predicted, leading to an increase in the expected failure rate. At least the following possibilities should be explored:

- Increase cooling provisions or reduce the ambient temperature (e.g., by relocating the circuit board).
- Reduce performance requirements to allow the unit to operate at lower speed, thereby reducing the internally generated heat.

- Switch to semiconductors that can operate at higher temperature.
- Accept the increased failure rate.

Once an engineering decision on the preferred approach is reached, contract administrators can assess the financial and schedule impact of this solution. Because of the interdependence of elements in a development schedule, it is important to become aware of difficulties (internal as well as suppliers') very early. This cannot be achieved when the parties engage in a cat-and-mouse game.

Design and evaluation tasks are listed in Table 8.2. As indicated in the comment column, many of these have been discussed in earlier chapters.

In addition to addressing the tasks in the table, the reliability plan should reference sources for design margins, derating provisions, and similar design guidelines relevant to reliability.

Reliability allocation can be functional or quantitative. The functional approach has been discussed in connection with Figure 8.4 in Section 8.2. The quantitative allocation involves parceling out the maximum allowable system failure probability to subsystems and major components. The purpose of the task is to establish early in the development that it is *possible* to achieve the required system reliability, and negative outcomes are much more significant than positive ones. For example, if the allocation requires that a transmitter

Table 8.2
Design and Evaluation Tasks

Description	Comments
Reliability modeling	See Chapter 2
Reliability allocation	See Section 8.3
Reliability predictions	See Chapters 2 and 4
Failure modes, effects, and criticality analysis	See Chapter 4
Sneak circuit analysis	See Chapter 4
Electronic parts/circuits tolerance analysis	See Section 8.3
Parts program	See Section 8.3
Reliability critical items	See Section 8.2
Effects of testing, storage, handling, packaging, and maintenance	See Section 8.3

failure rate be not more than 0.1 per 10^6 hours and the actual failure experience for this type of component is 2 per 10^6 hours, it is highly unlikely that the allocated failure rate can be achieved. On the other hand, an actual failure experience of 0.1 per 10^6 hours indicates that it may be possible to achieve the allocation but it does not ensure it. The amount of effort allocated to reliability allocation must be commensurate with this limitation.

In many current systems the reliability requirement is not expressed as a single failure rate but is stated in terms of specific failure modes. An example from a communication system is shown in Table 8.3. It is seen that the allocation by failure modes is much more suitable in this case than a single failure rate.

Electronic parts tolerance analysis has been incorporated into many of the circuit design tools that are routinely used today. Therefore it is now rarely considered a task to be conducted by reliability personnel. However, confirmation that the analysis was conducted and adequate margins were maintained should be obtained as part of the development phase reviews (see Section 8.4).

The need for a parts program arises when the developer (or customer) does not have a qualified supplier list or where a specified part cannot be obtained from a qualified supplier. Because of the high cost of establishing a low failure rate purely by test (see Chapter 5), a combination of design margin analysis and

Table 8.3
Reliability Requirements for a Telephone Switching System

Frequency of Service Impairment	**Objective**	**Units**
Probability of call cut-off	0.000125	(ratio)
Probability of ineffective machine attempts	0.0005	(ratio)
Line, trunk failure rate	15×10^{-6}	per hour
System failure rate	15×10^{-6}	per hour
Duration of Service Impairment		
Line, trunk downtime	28	min/yr
Partial system downtime	*	min/yr
Digital termination	20	min/yr
Total system downtime	3	min/yr

*varies as a function of system size
(*Adapted from:* [11].)

environmental testing may be employed. The responsibility for the parts program may be assigned either to the reliability or quality assurance function.

In many organizations adverse effects of testing, storage, handling, packaging, and maintenance are controlled by reference to industry standards or internal procedures. It is obvious that reliability is never improved by storage or other postdevelopment life-cycle activities; the aim is that it not be significantly degraded. Reliability engineers for military systems are particularly concerned with the effects of storage, which can extend more than a decade. Periodic testing is usually required to establish the operating status of stored weapon systems but there are indications that frequent testing has an adverse effect on reliability [12]. Trade-offs are therefore necessary to determine an optimum test frequency. Guidance on failure effects and failure rates due to storage can be found in [13].

The last of the major group of reliability tasks is concerned with testing. Four modalities of tests are included:

- Environmental stress screening (ESS) [14];
- Reliability development/growth test (RDGT);
- Reliability qualification test (RQT);
- Production reliability acceptance tests (PRAT).

Because of the high cost of reliability testing (see Chapter 5), these tests are practically never conducted at the system level. If they are conducted at the component level, it is important that they be motivated by a desire to identify failure mechanisms. The last three tests are described in [10]. They were originally targeted at random failures, mostly to demonstrate that a failure-rate specification has been met. They are no longer considered effective for that purpose for the following reasons:

1. They are usually conducted too late in the life cycle to reject a product—too much money has been spent already and it is usually difficult to find a replacement.
2. If the failures are truly random there is no easy way to reduce the failure rate (however, if they are concentrated in one or a small group of parts, remedial action may be possible).

8.4 Reviews and Audits

As used here, the term *review* means a technical evaluation of work accomplished (usually by the developer) that may be limited to a specific aspect of the system (e.g., software or interfaces) or to a development phase (requirements, design, and so on) but can go to any level of technical depth. An *audit* is concerned with the procedures that were used in a specific phase (e.g., that the specified standard for FMEA was followed or that the design tools incorporated an electronic parts tolerance analysis). The results of an audit are generally to accept a process or task or to reject it (requiring it to be repeated). A review allows a wider scope for failure prevention activities, including analyzing design margins, establishing requirements for tests to verify reliability-related attributes, or setting limits on the operational envelope. The emphasis in the following is on reviews.

Reliability reviews and audits are only rarely stand-alone events. They are typically part of system or product reviews and audits in which reliability plays an important part, but still only a part. These reviews are frequently scheduled at the end of a life-cycle phase but they can also be focused on an event such as completion of a test or delivery of a report. The former military standard for reviews and audits, MIL-STD-1521, was completely waterfall model-oriented [15]. The more recent IEEE Standard for Software Reviews, IEEE Std 1028, addresses primarily individual documents and events but also recognizes some life-cycle phases [16]. A comparison of the life-cycle phases between these two standards is shown in Table 8.4.

Reliability-relevant events or documents that are the subject of a review or audit in the IEEE standard include the following:

Table 8.4
Comparison of Review and Audit Standards

MIL-STD-1521	IEEE Std 1028	Remarks
System requirements review		
System design review		
Software specification review	Software requirements specification	
Preliminary design review (PDR)	Software design description	PDR in IEEE Std 730 [17]
Critical design review (CDR)	Source code review	CDR in IEEE Std 730
Test readiness review (TRR)	Software test documentation	TRR in IEEE Std 1012 [18]
Functional configuration audit		Functional audit in IEEE 730
Physical configuration audit		Physical audit in IEEE 730
Formal configuration review		SW configuration management review in IEEE 730
Production readiness review		

- Back-up and recovery plans;
- Customer or user complaints;
- Anomaly reports;
- Maintenance plans;
- Risk management plans.

First a few words about the format of a review and then a discussion on the reliability-relevant subject matter that is typically covered. Preparation is the single most important factor in making a review effective. Preparation is required of all participants but it is particularly essential for the lead personnel of each organization; let us call them the developer lead and the customer lead. The developer lead is responsible for assembling all the materials for the review, scheduling the sessions, and having knowledgeable presenters available. The developer lead is also usually charged with presenting contractual review objectives, furnishing the recorder, and keeping track of action items. The customer lead must be thoroughly familiar with the contractual review objectives, the project history (particularly holdover action items and pending change proposals), and he or she should have a prepared plan of action to deal with deviations that become evident. There is always a conflict between saving money, saving time, and achieving reliability objectives; if the customer lead is not prepared to resolve these conflicts it will delay problem resolution.

The first session of a review usually includes an introduction of the participants and a general review of the current state of the requirements or design. On major systems, subsequent sessions may be conducted in working groups organized by components or disciplines. If the latter approach is taken, it is desirable to cover reliability topics in conjunction with maintainability, testability, safety, and quality assurance because of the many interactions between these issues. Reliability personnel should also be involved in the discussions of testing (including test plans), training, and logistics. At the conclusion of the working group sessions, the group spokesperson reports the group's decisions to a plenary session. In less-complex systems the formation of working groups may not be necessary.

The results of a review are usually organized under two major headings: decisions and action items. Decisions can take the following form:

- Unconditional acceptance of the results and authorization to proceed into the next phase;
- Conditional acceptance, pending resolution of specified action items, and conditional authorization to proceed into the next phase (delaying some work until pertinent action items are completed);

- Continuation of the review until a specified date without authorization to proceed into the next phase;
- Declaring the results unsatisfactory, in which case there must be a substantial reformulation of the system or its abandonment.

The most frequent outcome is the second bullet, and that leads us to a discussion of action items.

Action items can be simple instructions, such as to add a section to a report, change a dimension, or add a redundant power supply. Other action items include requirements for additional analyses or trade-offs or to resolve interface problems, typically with other user equipment. The results of these action items have to be reviewed and that introduces schedule uncertainty. The customer (user) can reduce this by specifying the course of action for each outcome of the analysis or other activity.

Action items can also be imposed on the user. Typical reasons for this are:

- Incomplete specification of an interface;
- Referencing a standard without supplying parameters necessary for compliance;
- Deferring acceptance of a document pending review;
- Deferring a major design decision pending availability of funds.

Progress on a project can be significantly impeded by delays in responding to the action items.

The most frequently encountered review topics that concern system reliability are the following (topics encountered in specialized applications are discussed later):

- Statement of reliability requirements:
  - Adequacy of the format (e.g., compared to Table 8.3);
  - Verifiability of the requirements (see Chapter 5; very high reliability requirements may have to be verified by inference and the process for this should be identified);
  - Mandatory and guidance standards;
  - Identification of critical failure modes and reliability-critical items.
- Reliability plan:
  - Format and responsibility assignments;
  - Execution of plan;
  - Provisions for software reliability;

- Statement of design margin and other quality requirements in the plan and verification provisions for these.
- Failure rate information:
  - Suitability of sources, particularly whether these have led to valid reliability estimates in the developer's prior products;
  - Partitioning to part failure modes;
  - Responsibility for verification of critical failure rates.
- Preliminary or definitive component failure rate estimates:
  - Methodology and assumptions used for estimation;
  - Separation by usage phases (e.g., ground operations, take-off, cruise, and so on).
- Reliability analyses (see Chapter 4):
  - Methodology and assumptions;
  - Prior use of this methodology and validity of its results;
  - Reasonableness of current results and plans for their validation, particularly for critical items;
  - Integrated or separate analysis of software reliability;
  - Life analysis of parts subject to wear-out or depletion.
- Reliability modeling:
  - Modeling tools and assumptions;
  - Format of modeling results (usage phases, component aging);
  - Parameter verification for critical items (see Section 8.5);
  - Reasonableness of modeling and verification results versus requirements.
- Redundancy provisions (see Chapter 6):
  - Testability of individual channels prior to and in operation;
  - Monitoring of switching and reconfiguration;
  - Verification of software used for failure detection and reconfiguration.
- Availability:
  - Adequacy of model and assumptions;
  - Built-in test and off-line diagnostics;
  - Maintenance access;
  - Hot-switch capability for electronic modules;
  - Provisions for rapid changes in software.
- Supplier reliability efforts (see also Section 8.5):
  - Method of coordination and supervision;
  - Prior procurement from each supplier and reliability experience;
  - Results of supplier efforts to date.

- Reliability data from test programs:
  - Evaluation of testing and inspections for reliability parameters;
  - Plans for further testing and evaluation.
- Failure reporting system:
  - Format, responsibility for its operation, and current status;
  - Failure review board;
  - Summary of reliability problems identified;
  - Discussion of unresolved reliability problems;
  - Statistics of current operations:
    - Total number of reports received;
    - Number of reports rejected;
    - Number of reports closed out;
    - Number of reports pending.

Topics that arise in specialized application environments include the following:

- Radiation effects on parts and component reliability;
- Safety and reliability demonstrations required for airworthiness certificates;
- Source control of reliability-critical items (assurance that no aspects of these items are changed without knowledge of and approval by the user);
- Effects of the space environment on parts and component reliability (radiation and electromagnetic fields, meteoroids);
- Effects of tooling, manufacturing processes, and shipping for mass-produced items;
- Effects of long-term storage, particularly for military systems.

8.5 Monitoring of Critical Items

Items that are critical to safety, reliability, or availability should be monitored during development and production for (a) adherence to the specified process, and (b) compliance with performance, quality, and reliability requirements. In the following we review the objectives and some of the practices of monitoring purchased items and items produced in-house.

8.5.1 Monitoring Purchased Items

In the procurement of critical items for incorporation into products, the purchaser usually has considerable leverage in obtaining data and support agreements from the vendor. To safeguard the reliability of the end item, the following subjects should be brought up for discussion:

- Configuration control;
- Notification of deviations from internal process control;
- Notification of adverse experience by other users, suppliers, and distributors.

These subjects will now be discussed in some detail. The inspection of incoming items is covered under the in-house monitoring subjects.

Configuration Control

Configuration control provides assurance that all items procured under a given contract will be produced to the same drawings and processes. Configuration control is particularly important when reliability estimates are based on experience with a particular design. Anecdotes abound about "improvements" introduced by well-meaning vendors that have caused reliability problems when the "improved" parts were used in an environment that the vendor did not understand.

Configuration control should be a *flow-down* requirement, meaning that lower tier suppliers of parts and process items (cleaning solutions, adhesives, and so on) should also be bound by the provisions. When there are compelling reasons to make changes, these must communicated and samples of the changed product furnished for evaluation. Marking with a serial number or lot number can help identify components that were produced during an inadvertent deviation from the established configuration control procedures.

The configuration control agreement should obligate the vendor to give advance notice of intended product improvements. The advance notification period should be sufficient to allow for evaluation of the proposed improvement and possible selection of another vendor.

Process Control Deviations

Manufacturers maintain control charts or similar data on key parameters of the incoming materials and on intermediate process steps and final test. Deviations from expected patterns of the control charts usually cause an adjustment in the process and retraining or reassignment of personnel. Although the finished items produced during the time that the deviation existed still meet all functional specifications, they may have lower reliability (e.g. with respect to protective coatings). Vendors usually do not give nongovernment customers access to their internal quality records but agreements can sometimes be obtained for notification of adjustments in process steps that are particularly critical for the intended application.

Adverse Experience by Other Users

Adverse reliability experience by other users of a product can provide valuable warning of impending reliability problems for the intended application. Procurement documents should obligate the vendor to immediate notification if adverse events become known. Not every problem encountered by another user is necessarily relevant because the problem may have been due to unusual environmental conditions, abuse by operators, or exceeding specified operating parameters. But an investigation of the relevance of these problems is always warranted.

The notification of adverse experience should also be a flow-down requirement, applicable to suppliers and distributors of the vendor. A valuable source of information about adverse experience on widely used parts is the Government-Industry Data Exchange Program (GIDEP), a cooperative activity between government and industry participants seeking to reduce or eliminate expenditures of resources by sharing technical information. The GIDEP Web site is maintained by the Corona Division of the Navy Surface Warfare Center.

8.5.2 In-House Monitoring for Reliability Attainment

All the provisions that were found to be desirable for suppliers are also desirable for the in-house development effort: configuration control, monitoring of process control, and awareness of adverse experience by others. The responsibility for configuration control and monitoring of process control are usually assigned to departments other than reliability but reliability staff must identify reliability concerns for components and their critical characteristics to these departments. It is also essential that reliability personnel understand the techniques used by quality assurance and parts specialists. To aid in these efforts, we provide details on the following:

- Quality assurance;
- Failure reporting and analysis;
- Test specifications, reports, and analysis.

Quality Assurance

One frequently encountered tool of quality assurance is the control chart, in which the value of a critical parameter is plotted in serial fashion for each inspected unit. The following discussion applies to incoming inspection of purchased items, as well as to monitoring in-house production. Figure 8.6(a) shows a control chart for a well-controlled process. In Figure 8.6(b) we can see an abrupt reduction in the parameter between unit 4 and unit 5 that persists for all subsequent units. This pattern is typical of a change in the setpoint of an otherwise well-controlled process. This change in setpoint of the process (or the instrumentation) should be investigated for reliability implications.

In Figure 8.6(c) there is an apparent upward slope in parameter values without an abrupt change. The average parameter value of the last five units is about 0.1 greater than that of the first five. This pattern is indicative of a drift in the setpoint that must be immediately brought to the attention of manufacturing personnel to prevent production of units that have functional or reliability deficiencies. Figure 8.6(d) shows a process that is clearly out of control and cannot be depended on to produce reliable equipment.

Control charts aid in visualizing what is happening to a process but when hundreds of parameters have to be reviewed, the documentation can become unwieldy. Presentation of simple statistical parameters can identify processes that may be in trouble and for which control charts and other documents may

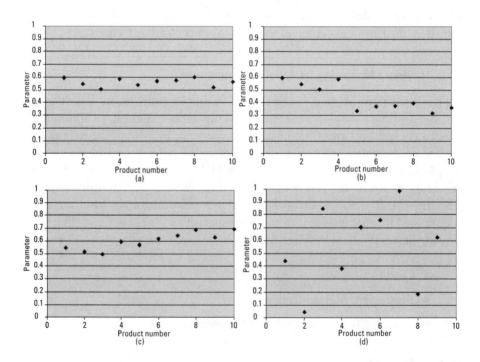

Figure 8.6 Use of control charts: (a) satisfactory control, (b) change in setpoint, (c) drift of setpoint, and (d) loss of control.

have to be examined. Table 8.5 presents the average and standard deviation for each of the charts shown in Figure 8.6. Assuming that reference data for a process under satisfactory control, such as Figure 8.6(a), is available, each of the other data sets would be flagged as probably resulting from a process deviation.

The data for Figure 8.6(b) would be flagged because the average is approximately five standard deviations lower than that of the reference process and the standard deviation is five times greater (each of these differences by themselves indicates the need for further investigation). For Figure 8.6(c), the average is two standard deviations higher than the references and the standard deviation is almost three times that of the reference. Here the simultaneous occurrence of two moderately significant differences invites further investigation. For Figure 8.6(d), the tenfold increase in standard deviation clearly signals that the process has gone out of control.

Failure Reporting and Analysis

The format of failure reports has been discussed. Here we concentrate on the use and analysis of failure reports during development and early production. One question that arises frequently is when to start a formal failure reporting system. An argument against an early start is that the design should be stable so that the database does not include failures due to features that have been changed because they were recognized as unreliable. Inclusion of these failures can lead to misleading results about failure rates. Our suggestion is that reports pertaining to failure mechanisms that have been removed (regardless of what stage of the development cycle) be tagged so that they can be omitted from summary statistics. Once this is done, there are very few arguments against starting failure reporting as soon as tests on incoming materials and laboratory testing become available.

The review of failure reports by reliability staff should accomplish at least two objectives: identify the immediate cause of the failure and determine whether the same or a related cause may be present in other areas of the design. The latter activity is sometimes referred to as *root cause analysis* (see also Figure 8.5).

Table 8.5
Statistical Data for Four Processes

Data Item	Figure 8.6(a)	Figure 8.6(b)	Figure 8.6(c)	Figure 8.6(d)
Average	0.554	0.434	0.594	0.542
Standard deviation	0.022	0.112	0.064	0.215

The immediate cause can usually be determined by the designer or by a task group composed of the designer, a parts or materials specialist, and the reliability engineer. Examples of immediate causes are: a high-resistance solder joint, a shorted capacitor, or a high threshold voltage in a switching transistor.

The high-resistance solder joint will be examined for visual cues of a "cold joint" condition. If these are found, insufficient inspection will be noted as a contributory immediate cause. Possible root causes include the following:

- Poor process control for soldering;
- Mechanical damage to board due to lack of protective covering;
- Excessive heat that could have weakened the solder joints.

In general, for each identified immediate cause, two questions must be raised:

1. Could this defect have been detected at an earlier stage of assembly?
2. What change in procedures would have prevented this defect from occurring in the first place?

The effort put into these investigations is governed by the history of similar failures, the cost of the in-process failures, and the potential cost of such failures that may occur after delivery. The impact of in-service failures on the company's reputation and regulatory or legal liabilities (such as mandatory recalls) must also be considered.

Awareness of Technical Information

Awareness of current information on reliability trends, experience with specific parts and processes, and failure detection and prevention practices is a primary responsibility of the reliability engineering staff. Membership in a professional society that provides access to their publications, conferences, and restricted Web sites is an important step toward that objective. Local meetings sponsored by the societies permit informal exchanges that can either provide needed information directly or point to venues where it can be obtained. Organizational membership in GIDEP (see Section 8.5.1) provides another source of technical information.

Web sites give access to current manufacturer's data that should be perused to avoid improper application of purchased components. Application notes or service bulletins posted on the Web sites may describe general precautions necessary to avoid reliability problems and sometimes discuss specific failure modes such as damage from electrostatic discharge.

8.6 Chapter Summary

Causes of failure can be introduced at any phase of the life cycle, and therefore failure prevention activities should be undertaken in each phase. To motivate and guide these failure prevention activities, we examined life-cycle concepts and models. Because requirements for large systems are likely to change as development proceeds, life-cycle models and concepts that incorporate this need for change are now favored. Prominent among these are the spiral development model and rapid prototyping.

The major contributions to failure prevention that can be expected from the concept phase deal with a framework from which development phase activities can proceed. The concept phase should identify:

- The adopted life-cycle model and dates for each major phase;
- Reliability stakeholders (users, regulatory agencies, and so on) and their expectations;
- Special reliability requirements (to prevent critical failure modes);
- Experience from similar prior systems.

The special requirements and failure experience from prior systems combined point to failure modes that require the greatest attention and around which the development phase activities should be planned.

The bulk of the reliability activities occur during the development phase, and they include all of the topic areas that have been presented in previous chapters of this book: reliability estimation and modeling, recognition of how systems fail, reliability analysis, testing to identify failure mechanisms, redundancy, and the contribution of software. The initial planning for the development phase should center on prevention of critical failure modes, later shifting to satisfying the formal reliability requirements as they evolve.

The FRACAS is the focus of reliability activities in the O&M phase. Every failure report is an opportunity to learn. It will not only identify the cause of a specific failure but can also point to failure patterns, specific components that undergo many repairs (repeaters), and may show unrecognized failure modes in the FMEA.

Suggestions for a formal reliability program plan are presented in Section 8.3. The extent and formality of this plan depend on the size of the system and the management style. But the reliability of complex systems requires contributions from many functions and individuals, and the plan can be an excellent vehicle for assigning responsibilities, scheduling activities, and making sure that all bases are covered.

In Section 8.4 we discussed the role of reviews and audits that are usually conducted at the end of development phases but can also be held to evaluate individual products, such as reports and demonstrations. The format of reviews is discussed and the need for preparation by all participants is emphasized. The section contains a structured list of review topics pertinent to reliability. These topics, together with discussions in earlier parts of this chapter, will permit the reader to become an effective participant in the reliability sessions of a program review.

The last topic section of the chapter deals with monitoring critical items during development and production. We note that the process must be controlled through configuration control, the product through control charts or equivalent quality-control practices, and that the reliability engineer must be watchful for any adverse experience with items or processes by other users.

References

[1] Blair, G. M., *Starting to Manage: The Essential Skills*, Piscataway NJ: IEEE Press, 1996.

[2] http://www.asprova.com/en/kaizen/words/gantt.html.

[3] Royce, W. W., "Managing the Development of Large Software Systems," *Proc. IEEE WESCON*, San Francisco, August 1970.

[4] http://asd-www.larc.nasa.gov/barkstrom/public.

[5] Boehm, B., "A Spiral Model of Software Development and Enhancement," *IEEE Computer*, May 1988, p. 61.

[6] Pressman, R. *Software Engineering: A Practitioner's Approach, European Edition*, New York: McGraw Hill, 1997.

[7] Beck, K., *Extreme Programming Explained*, Addison Wesley, September 1999.

[8] Bosnak, R. G., "Improving Reliability and Maintainability Programs With Configuration Management," *Proc. of the 1990 Reliability and Maintainability Symposium*, Los Angeles, CA, January 1990, pp. 109–112.

[9] Crow, L. H., P. H. Franklin, and N. B. Robbins, "Principles of Successful Reliability Growth Applications," *Proc. of the 1994 Reliability and Maintainability Symposium*, Anaheim, CA, January 1994, pp. 157–161.

[10] U.S. Department of Defense, MIL-STD-785B, *Reliability Program for Systems and Equipment, Development and Production*, August 1988, canceled 1998 but copies still available from ASSIST, a service of the DoD Document Automation and Production Service, 700 Robbins Avenue, Philadelphia, PA, 19111.

[11] Fletcher, L. A., D. E. Burns, and W. P. Cochrane, "Motivating Reliable Switching System Performance," *Proc. of the 1989 Reliability and Maintainability Symposium*, Atlanta, GA, January 1989, pp. 383–388.

[12] Rooney, J. P., "Storage Reliability," *Proc. of the 1989 Reliability and Maintainability Symposium,* Atlanta, GA, January 1989, pp. 178–182.

[13] Rossi, M. J., *Nonoperating Reliability Databook,* Utica, NY: Reliability Analysis Center Publication NONOP-11987.

[14] START (Selected Topics in Assurance Related Technologies), *Environmental Stress Screening,* Vol. 7, No. 3, Utica, NY: Reliability Analysis Center, 2000.

[15] U. S. Department of Defense, *Technical Reviews and Audits for Systems, Equipments, and Computer Software,* canceled 1995 but copies still available from ASSIST, a service of the DoD Document Automation and Production Service, 700 Robbins Avenue, Philadelphia, PA, 19111.

[16] Institute of Electrical and Electronics Engineers, *IEEE Standard for Software Reviews,* IEEE Std 1028, New York, 1996.

[17] Institute of Electrical and Electronics Engineers, *IEEE Standard for Software Quality Assurance Plans,* IEEE Std 730, New York, 1989.

[18] Institute of Electrical and Electronics Engineers, *IEEE Standard for Software Verification and Validation Plans,* IEEE Std 1012, New York, 1986.

9

Cost of Failure and Failure Prevention

The bulk of reliability literature is devoted to activities for meeting a specified reliability requirement or evaluating the progress toward meeting it. Chapters 2 through 8 also fall into this category. However, at this point we will deviate from this pattern and address the question "How do you set a reliability requirement for a new or modified component or system?"

To introduce the principles used to answer this question, we consider equipment that is not subject to regulatory reliability oversight and where there is no interaction of failure modes. The concept of optimum reliability for such an item is developed in Section 9.1. The time relationship of the expenditures for cost of failure and failure prevention are discussed in Section 9.2, and practical guidance for estimating the cost elements is presented in Section 9.3. General rules for the cost of reliability improvement are developed in Section 9.4, followed by the chapter summary.

9.1 Optimum Reliability

The concept of optimum reliability sounds strange, perhaps even perverse. Aren't developers motivated to make their product as reliable as possible? Of course, but they are constrained by the need to keep the product competitive in price and for getting it out on schedule. The costs and time required for failure prevention are concrete and must be faced immediately; the consequences of economizing in this area are much less concrete and occur in the future, if at all. In the following we look at concepts that can lead to a rational decision between the conflicting requirements of designing a product for lowest cost and preventing it from failing in service.

The starting point is a low-cost product and its possibly unacceptable failure rate. The product's cost is arbitrarily assigned a value of zero, and we estimate its failure probability by methods discussed in Chapter 2 or based on that of other products with which the developer is familiar. To keep the analysis simple we assume that the item will be used only over a short interval so that the time of the possible failure does not have to be varied. The event of failure will impose a loss on the user, to which we assign the symbol C_f. The expected value of this cost of failure is

$$\mathrm{E}[V_f] = f \times C_f \qquad (9.1)$$

where f is the probability of failure.[1] Improving the quality or robustness of the product can reduce the failure probability, but that reliability improvement will increase the cost to the user by an amount ΔV_r. From the user's point of view, a given reliability improvement will be desirable as long as

$$-\Delta \mathrm{E}[V_f] \geq \Delta V_r \qquad (9.2)$$

The cost increment for reliability will be considered beneficial if it is not more than the decrement in the cost of failure. The relationship between expenditures for reliability and avoidance of cost of failure can also be explored by postulating that both cost elements must ultimately be borne by the user. If the user's expenditures are designated as V_u we have

$$V_u = V_r + \mathrm{E}[V_f] \qquad (9.3)$$

It is understood that improvement decisions always involve increments and we have therefore dropped the Δ symbols. The user's total cost and its two constituents are graphed in Figure 9.1. The values shown for failure probability and cost increment are arbitrary. The cost of failure rises linearly with failure probability, as is evident from (9.1). A long-established figure of merit for reliability improvement has been failure probability removed per unit expenditure [1]. A decision maker using this criterion will first institute the improvements that have the shallowest slope (right end of the cost of reliability curve in Figure 9.1) and then consider those with a successively steeper slope. Thus, at the low failure probability side of the graph, only the costliest improvement

---

1. The implied linear relationship between probability of failure and cost of failure will be challenged later in this chapter.

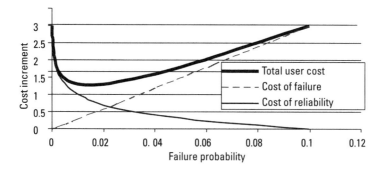

Figure 9.1 Cost relations for reliability improvement.

opportunities remain, resulting in the very steep slope. Zero failure probability can be approached but not reached.

The most significant aspect of this graph is that the total user cost has a minimum, and the failure probability at which the minimum is reached represents the optimum reliability under economic motivation. A lower failure probability can be achieved, but the cost for doing this will be higher than the reduction in the expected cost of failure [the inequality of (9.2) will be violated].

At the minimum, the slope of the cost of reliability curve is the negative of the slope of the cost of failure line. From (9.1)

$$\frac{d\mathrm{E}[V_f]}{df} = -C_f \tag{9.4}$$

The right hand side of the equation is a constant that is usually known for a given application. It follows that the reliability improvement alternatives that should be considered will be limited to those for which

$$-\frac{dV_r}{df} \leq C_f \tag{9.5}$$

While C_f is constant for a given application, it can vary between applications. Consider a strip heater that is sold to prevent freezing of exposed water pipes in a part of the country that is only occasionally subject to prolonged freezing temperatures. If it fails, the user will have to call a plumber, replace the heater, and possibly pay for water that leaked if the pipe broke. Let us be generous and throw in some dollars to compensate the user for the inconvenience and we arrive at C_f = $500. Now consider the same strip heater being used for an oil pipeline in Alaska. A failure will cause interruption of the oil supply, need for a repair crew, possible environmental cleanup liability. An estimate of $C_f =$ $1,000,000 may be on the low side. Taking it one step further, a strip heater may be used to prevent freezing of a propellant line on a spacecraft. Failure of the propellant line will cause loss of the spacecraft. The failure will require a replacement spacecraft and its launch, expenditures for failure analysis, and repair of public relations damage, plus the salaries of the spacecraft operations crew for an extended period of time. An estimate of $C_f = $50,000,000$ may not be excessive.

As C_f increases so does the slope of the cost of failure line and that, in turn, causes the minimum of the total user cost to move toward the left, toward lower failure probability. The cost of reliability curve is not affected by the changes in the application, but the move of the minimum toward lower failure probability requires higher expenditures for reliability to remain in the optimum region. There are practical implications to this analysis: the use of premium grade electronic parts for industrial and military applications, and the use of super grade (Class S) parts for space applications.

It will also be noted that the minimum of the total user cost curve is shallow. This means that there is a range of failure probability over which the total user cost is at approximately the same low value. Errors in estimating the parameters for the user cost (and its constituents) will therefore not invalidate the optimum reliability determination. In Figure 9.1 the total user cost is within 10% of the minimum for failure probabilities between 0.01 and 0.03, and within 20% of the minimum for failure probabilities between 0.005 and 0.04. While this range provides some comfort to the decision maker, it must also be noted that a 1% saving on a $10,000,000 reliability improvement represents $100,000.

9.2 Time Considerations of Expenditures

The cost for reliability improvement is always incurred before the expected cost of failure arises. This creates financial, psychological, and administrative problems. Most of this section is devoted to accounting for the financial effects of this interval between the cost and benefits of a reliability improvement program. But at the outset we want to acknowledge the psychological and administrative obstacles to expenditures for reliability.

Reliability improvement requires concrete expenditures for analysis, test, stronger or higher quality parts, redundancy, and failure monitoring provisions. The benefit is a reduction in the *expected* cost of failure—something much less concrete, measurable, and certain than the expenditures. Regardless of whether funding for reliability comes out of the developer's funds or is paid for by a sponsor (who represents the user), there is reluctance to incur concrete and current expenditures for an uncertain future benefit. The most important tool for dealing with this normal problem is data—test data, field data, supplier data, and reports and analyses by outside organizations. The most convincing source is usually warranty data because this indicates not only the frequency of failure but also the cost to the producer.

Administrative obstacles arise from the fact that expenditures for reliability may have to be borne by a different entity than the cost of failure, and sometimes reliability expenditures during development come from a different source than those during production. The classical demonstration of difficulties that can arise is the indifference of the U.S. automotive industry to the cost of failure borne by the consumer until foreign companies proved that U.S. consumers are willing to pay more for a reliable car. Much of the regulatory apparatus is intended to make producers responsible for keeping the cost of failure that must be borne by the user at a reasonable level. Product service warranties demanded by the military for many weapon systems are aimed at the same problem. Reliability practitioners need to be aware that there is product differentiation on the basis price—that users of a low-end product will be more willing to accept high failure probability than those of a high-end product. And they need to be mindful of the lessons learned by the automotive industry.

Turning now to the financial issues that arise from the time lapse between expenditures and benefits, we will first consider a project with a single expenditure for reliability improvement and a single benefit of reduced failure probability. The development of a single spacecraft and its subsequent launch fits this schema. The time history of resource flows is shown in Figure 9.2. Downward pointing arrows signify expenditures and upward pointing ones are benefits.

The diagram can be interpreted as a loan transaction for which we are now trying to assign a suitable rate of interest. At the time the reliability improvement is funded, the following information is usually available:

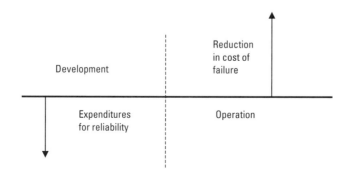

Figure 9.2 Single expenditure and benefit.

- The repayment date is an estimate with no certainty of it being achieved.
- The amount to be repaid is an estimate with no certainty of any amount.
- There is no signer, cosigner, or collateral.

These are hardly conditions for extending a loan at the prime interest rate. Depending on the experience with the intended improvement and the scheduling of the benefit period, interest rates of 10% to 20% annually will be reasonable. Using annual compounding, the required benefit (at the end of the period) for each dollar of reliability improvement funding is shown in Table 9.1. The amounts are computed from

$$V'_f = (1+i)^n \tag{9.6}$$

where V'_f is the decrement in the cost of failure per $1 reliability improvement resource, i is the fractional interest rate, and n is the number of years between expenditure and benefit.

As can be seen in the table, higher interest assumptions and a longer time to benefit can significantly increase the amount of benefit required to warrant a given expenditure. This is a fact of life that must be faced in any economically motivated reliability improvement program.

In a more frequently encountered situation the expenditures as well as benefits are spread over a number of years. We start with the simplest case, in which the expenditures are the same in each period and the benefits are also uniform, as diagrammed in Figure 9.3. The figure assumes that the expenditures occur during development, there is a latency period (representing preproduction and production) before the product is placed into service, and then there is an

Table 9.1
Required Benefit to Justify $1 Reliability Improvement Expenditure

Years to Benefit	Annual Interest		
	0.1 (10%)	0.15 (15%)	0.2 (20%)
1	1.10	1.15	1.20
2	1.21	1.32	1.44
3	1.33	1.52	1.73
4	1.46	1.75	2.07
5	1.61	2.01	2.49

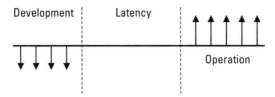

Figure 9.3 Periodic uniform expenditures and benefits.

operation period during which the benefits of reduced failure probability are achieved.

The following is the general three-step procedure for the financial modeling of these expenditure and income streams:

1. Convert the expenditures during the development period into a future value at the end of the development period, FV_1.
2. Account for the latency period by the compound interest model (9.6). Multiplying FV_1 by $(1 + i)^n$ will result in the future value at the beginning of the operation period, FV_2.
3. Convert FV_2 into a uniform series of payments that constitute the minimum benefit due to failure reduction to warrant the reliability improvement expenditure.

The future value (FV) of a uniform series of $1 payments (step 1) can be computed from

$$FV = \frac{(1+i)^n - 1}{i} \qquad (9.7)$$

where the symbols have been previously defined. Microsoft Excel™ performs this calculation as the function FV. The series of end-of-period uniform payments corresponding to a $1 investment at the beginning of the period (step 3) is given by

$$PMT = \frac{i}{1-(1+i)^{-n}} \qquad (9.8)$$

and can be calculated by Microsoft Excel as the function PMT.

We apply the three-step procedure for the following conditions:

Annual interest rate (i) 0.15

Number of expenditure periods (n)   4
Amount of each expenditure $1
Number of latency periods   2
Number of benefit periods (n)   5

And we want to determine the amount of each benefit

Step 1. Apply (9.7) with $i = 0.15$ and $n = 4$. Result $FV_1 = 4.99$.

Step 2. Compute V'_f for $i = 0.15$ and $n = 2$, using (9.6), which yields 1.32 (see also Table 9.1). Multiply the result of step 1 by this value and obtain $FV_2 = 6.59$.

Step 3. Apply (9.8) with $i = 0.15$ and $n = 5$. Result: $PMT = 0.23$. Multiply this by FV_2 and obtain $6.59 \times 0.23 = 1.52$.

Thus, the amount of each benefit will have to be at least $1.52 to justify the expenditure of $1 during each of the development periods.

Where expenditures or benefits are not uniform, the amounts for individual periods have to be transformed to a common time reference by use of (9.6). The most convenient time reference is usually the start of the benefit period.

9.3 Estimation of Cost Elements

In this section we discuss estimating the cost and benefits when a specific reliability improvement is to be evaluated. Generic cost estimation relationships are developed in Section 9.4. We consider the following four general classes of reliability improvements:

1. *Process improvement*—usually effective only against a very narrow spectrum of failure mechanisms (e.g., attachment failures, contamination of semiconductors, voids in encapsulated components). This original process may have been faulty and the improvement is really a correction of these faults. The product cost increment is usually very small and can be neglected. But start-up costs can be significant. The introduction of a new process may also delay the overall schedule and interrupt the income stream.

2. *Increased design margins*—applicable to a wider but still limited number of failure mechanisms. Examples include derating of electronic and electrical parts, increasing the dimensions of mechanical parts, and providing better temperature or vibration control. These measures usually have only a minor effect on start-up costs but increase production cost not only by the higher purchase price of the affected

parts but also by the increase in weight, volume, and sometimes power consumption.

3. *Screening*—can be applied to parts or subassemblies but is effective primarily against known failures modes. Just as process improvement is frequently a fix for process defects, screening is intended to overcome insufficient controls in manufacturing or assembly. There can be a significant start-up cost, and there will be recurring costs due to the testing and cost of rejected product.

4. *Redundancy*—effective against a broad spectrum of failure mechanisms and most effective against random failures. Thus it complements the previously mentioned modalities that primarily addressed known (nonrandom) failure modes. Redundancy is most effective when employed at the assembly or higher level. Redundancy requires little expenditure during the development phase but large expenditures in the cost of the product. As discussed under number 2, trade-offs for the increase in weight, volume, and power will be required. In addition, redundancy carries with it a need for additional maintenance during the operations phase (more units are subject to failure).

The expected cost of failure depends on two factors [see (9.1)]: the failure probability and the loss caused by the failure. The same factors enter into the evaluation of the reduction in the cost of failure that is the benefit part of our considerations. Table 9.2 lists factors pertinent to estimating the failure probability for each of the four improvement classifications. The relative cost of the last three improvement methodologies is discussed in [2].

The main component of risk for redundancy is that systematic failure causes might have been overlooked. The estimation risk for this improvement is

Table 9.2
Estimating the Reduction in Failure Probability

Improvement	Failure Causes Eliminated	Estimation Risk
Process	Due to process deficiencies; reduction can usually be estimated from failure analyses and reports	Medium
Design margin	Due to underdesign or overstress; usually difficult to distinguish from random failures in failure reports	High
Screening	Due to lack of part or process controls; usually difficult to distinguish from random failures in failure reports	High
Redundancy	Random failures; estimate from failure reports	Low

therefore classed as low. Process deficiencies usually become known from high in-process rejection rates and test failures. Failed product can be analyzed and the cause of the failure can be determined. The components of estimation risk are that the improvement may not be completely effective and that a high failure rate due to the process deficiency may have masked other failure mechanisms. The risk is therefore classified as medium. Failures due to the other two improvement methods are usually recognized only from field failure data, and product may not be available for detailed failure analysis. The reduction in failure probability due to increased design margins and screening is therefore difficult to estimate and the estimation risk is designated as high.

The cost of failure prevention can be obtained from vendor quotes (e.g., substitution of a higher rated component for a nominally rated one, addition of a redundant power supply) or estimates of labor cost (e.g., inspection or test time, failure review procedures).

We developed a general procedure for estimating the cost of failure prevention (reliability) where the preventive measure is redundancy, which is described in Section 9.4. The procedure was used to generate the cost of reliability curve in Figure 9.1. The redundancy model used to construct the curve serves as an upper limit when reliability improvements other than redundancy are employed. Note that the other modalities prevent a narrower spectrum of failures. The allocation of resources for failure prevention should aim at the minimum point on the user cost curve, here at a failure probability of about 0.015. Only a small increase in user cost will result from modest errors in these estimates because of the shallow minimum. The procedure's main purpose is to avoid making big mistakes in setting reliability goals either too high or too low.

The *expected cost of failure* is more difficult to evaluate. In principle it is the product of the *probability of failure* and the *loss* that results as a consequence of the failure. Also, in order to assign the correct discount factor (see Section 9.2) we must estimate the times at which the failures will occur. The following can reduce the uncertainties in failure probability estimates:

- Use of experience on predecessor systems;
- Vendor data;
- Public databases and publications.

This data must be adjusted for the specifics of the intended application, such as the following:

- Changes in mission duration;
- Changes in environments (temperature, vibration);
- New operating modes.

Figure 9.1 and earlier discussion of the cost of failure assume that cost is a linear function of failure probability. This is generally true if the failure frequency observed in operation is within the range tolerated by the user. The cost of failure will steeply increase if the failure frequency exceeds the tolerable range, often leading to abandonment of a product. Examples are notoriously unreliable car or tire models, some mobile phone services, and many software programs. Typically the excessive failure rates are not due to random failures but rather to deterministic ones stemming from skimpy design margins, lack of quality control, and, particularly for software, inadequate testing. To avoid getting into that nonlinear portion of the cost of failure graph it is necessary to take corrective action immediately once a deterministic failure mechanism has been identified.

9.4 A Generic Cost of Reliability Model

The generic model discussed in this section will be found particularly useful during concept generation and early system design for major projects. At that time decisions have to be made by a small staff that cannot undertake detailed studies and trade-offs. The failure prevention method that is most generally applicable, and is particularly effective against random failures, is redundancy. This section uses more mathematical notation than most of the chapters in this book. Readers who would rather not struggle with those parts may just want to look at (9.9) and (9.10) to become familiar with the notation and then use the graphical results of the other equations as presented in Figures 9.5 and 9.6. The allocation procedure presented at the end of the section is deliberately non-mathematical so as to be available to all readers.

The reliability of an *n-redundant* component was calculated in Chapter 2 as (2.4). This equation will now be interpreted for a partially redundant system, as shown in Figure 9.4, where each of the redundant parts, R1 and R2, has the same failure probability. The fraction of replication (measured by total resources such as cost and weight) is designated by n such that $n = 1$ represents a single string and $n = 2$ represents the entire system duplicated. Then the right part of (2.4) simplifies to

$$F = F_o^n \tag{9.9}$$

Here F_o designates the failure probability before the introduction of redundancy and F is the failure probability of the partially redundant configuration.

The amount of resources required to achieve this lower failure level is

$$V = nV_o \tag{9.10}$$

where V represents the resources after redundancy and V_o represents the resources for the original nonredundant arrangement. We use the broad term

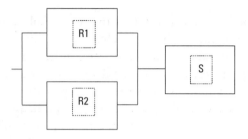

Figure 9.4 Partial redundancy.

resources and the symbol V (value) to account for weight and power consumption that are sometimes more critical than cost. But in most cases, particularly for electronic items, the cost is the governing resource and the only one that needs to be considered in initial planning. The *added* resource for redundancy is

$$v = (n-1)V_o \tag{9.11}$$

and the fractional increase, which will be designated u, is given by

$$u = v/V_o = n - 1 \tag{9.12}$$

For the partially redundant configuration shown in Figure 9.4, the value of n will lie between 0 (when only the S block is present) and 1 (when only the R block is present). For the notation introduced in (9.12), (9.9) can be rewritten as

$$F = F_o^{u+1} \tag{9.13}$$

This relation was used to obtain the cost of reliability curves in Figure 9.1. The marginal decrease in failure probability that can be achieved at a given level of resource expenditure can be computed from

$$\frac{dF}{du} = F_o^{u+1} \log F_o = F \log F_o \tag{9.14}$$

The results of (9.13) and (9.14) are plotted in Figure 9.5 for a nonredundant failure probability $F_o = 0.03$. Because dF/du is negative, we plotted the corresponding increase in reliability, labeled dR in the figure. Also, to make the plot more legible against the scale used for the failure probability, the value of dR has been multiplied by four. Figure 9.6 shows the reduction in failure probability for other values of F_o.

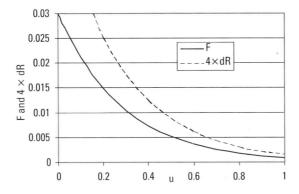

Figure 9.5 Failure reduction and marginal reliability increase for $F_o = 0.03$.

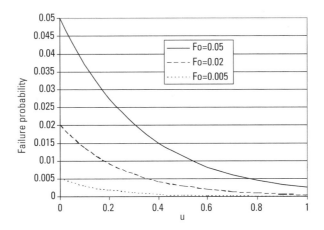

Figure 9.6 Reduction in failure probability for several values of F_o.

Thus, we have obtained tools for quantifying the increase in reliability that can be obtained by redundancy for a given resource expenditure. Figure 9.5 shows that the marginal reliability increment for a given expenditure, dR, decreases rapidly as the failure probability is reduced. The same tendency is also shown for discrete values of initial failure probability in Figure 9.6. This insight, together with reformulations explained next, can be the basis of an informed allocation strategy.

When $n = u+1$ [from (9.12)] is substituted in (9.10) and the latter is then differentiated, we obtain

$$\frac{dV}{du} = V_o = \frac{V}{u+1} \tag{9.15}$$

Combining (9.14) and (9.15) yields

$$\frac{dF}{dV} = \frac{F}{V}(u+1)\log F_o = \frac{F}{V}\log F_o^{u+1} = \frac{F}{V}\log F \qquad (9.16)$$

where the last form is obtained through use of (9.14). Representative values of $\log F$ are shown in Table 9.3.

The quantity dF/dV is indicative of the failure reduction that can be accomplished per unit resource expenditure, and therefore a high value of dF/dV is desirable. Equation (9.16) tells us that among items of approximately equal reliability, the greatest improvement per unit resource can be found among those with the highest failure/value (F/V) ratio. The statement also holds with respect to subsystems or components. Thus, in allocating resources among subsystems, we should first look at subsystems with the highest initial F/V ratio. Because the incremental reduction of failure probability per unit resource $\Delta F/\Delta V$ can be approximated by dF/dV, (9.16), together with Table 9.3, permits generic estimation of the benefits of a proposed reliability improvement.

In the typical allocation environment there is a fixed budget for reliability improvement and frequently also a reliability goal, but the responsibility for reconciling the two is nebulous. We assume that the budget is governing and develop an allocation procedure for that case. Equation (9.16) tells us that initially the most cost-effective reliability improvement can be obtained by redundancy of the component with the highest F/V ratio. At that point, we recalculate its F/V ratio, and if it has dropped below that of the (initially) second highest component, we introduce redundancy for that one. In the rare case where the first component still has the highest F/V ratio even after being dual-redundant,

Table 9.3
Values of $\log F$ for Typical System Reliabilities

Reliability, R	Failure Probability, F	logF
0.90	0.10	−2.30
0.93	0.07	−2.66
0.95	0.05	−3.00
0.97	0.03	−3.51
0.98	0.02	−3.91
0.99	0.01	−4.61

it may be considered for triple redundancy and that improvement is then compared to the benefit from making the second highest F/V component dual-redundant. The process is continued until either the budget is exhausted or the reliability goal has been achieved.

9.5 Chapter Summary

Most reliability activities are aimed at meeting a specified reliability requirement. In this chapter we introduced the question "How do you set a reliability requirement for a new or modified component or system?"

While the goal of every reliability organization is to make its product as reliable as possible, it is realized that the product must also be competitive in price and produced on schedule. The costs and time required for failure prevention are concrete and must be faced immediately; the consequences of economizing in this area are much less concrete and occur in the future, if at all. Thus, in Section 9.1 we looked at concepts that can lead to a rational decision between the conflicting requirements of designing a product for low cost and avoiding high failure rates in service. We concluded that the cost of reliability improvement must not exceed the expected cost of failures prevented by that improvement. Thus, we can speak of an economically optimum level of reliability at which incremental reliability expenditures will produce incremental reductions in the cost of failure that are exactly equal.

The cost for reliability improvement is always incurred before the expected cost of failure arises. This creates financial, psychological, and administrative problems. Most of Section 9.2 is devoted to accounting for the financial effects of this interval between the cost and benefits of a reliability improvement program. But we also speak briefly about the psychological and administrative factors that affect budgeting for reliability; particularly that the cost of failure is frequently borne by the consumer, whereas the cost of reliability improvement is borne by the producer.

Section 9.3 covers estimating the cost and benefits when a specific reliability improvement is to be evaluated. Four general classes of reliability improvements are discussed:

Process improvement—usually undertaken when the original process may have been faulty and the improvement is really a correction of these faults. The significant cost increments are start-up costs and schedule delays. Recurring costs are negligible.

Increased design margins—derating of electronic and electrical parts, increasing the dimensions of mechanical parts, and providing better temperature or vibration control. These measures increase recurring production costs as well as weight and volume. Start-up costs are usually small.

Screening—effective primarily against known failure modes in parts or subassemblies. There can be a significant start-up cost, and there will be recurring costs due to the testing and the cost of rejected product.

Redundancy—effective against a broad spectrum of failure mechanisms, and most effective against random failures. It increases cost, weight, volume, and power consumption and adds to maintenance cost during operation.

In estimating the expected cost of failure we must be aware of possible nonlinear effects that arise when systematic failures cause product recall and loss of consumer confidence.

The generic model discussed in Section 9.4 will be particularly useful during concept generation and early system design for major projects. The model is the basis for the cost of reliability curves used in earlier sections to define optimum reliability and motivate cost trade-offs. The F/V ratio (failure probability divided by the total resources devoted to a component) is a significant indicator for selecting components or subsystems for improvement.

References

[1] Hill, J., and J. D. Wells, "An Effective Index for Reliability Improvement," *Annals of the Assurance Sciences,* New York: ASME, 1965.

[2] Hecht, H., "Economic Factors in Planning Predictive Tests," *Material Research and Standards,* Vol. 12, No. 8, Philadelphia, PA: ASTM, 1972, pp. 19–24.

10

Cost Trade-offs

Chapter 9 provided the basic concepts for the economic evaluation of alternative reliability improvements. In this chapter these concepts will be applied to case studies covering a range of applications. Also, while Chapter 9 primarily addressed the needs of the concept phase when little specific design information is available, we will now discuss later life-cycle stages with more data but also more constraints.

In the first two examples we are part of the reliability staff of a communication company that has entered into a quality of service (QoS) agreement with its major customers. Our assignment is to recommend the reliability improvements that will result in the lowest total cost (improvement plus expected penalty).

Our fame has spread and we have accepted a higher paying job at one of the largest pharmaceutical manufacturers. This time we have been asked to evaluate the best way to reduce maintenance costs.

Finally we are employed by an operator of communications satellites (we really aim high) and have to decide when to launch a replacement satellite. This example involves a combination of reliability and logistic considerations because one of the drivers for replacement is the possible exhaustion of station-keeping fuel on the current satellites.

10.1 Reliability Improvement to Meet QoS Requirements

A company providing Web hosting for a number of large distributors has been pressured by a major customer to enter into a service agreement with the following provisions for the service as a whole (service for individual lines is covered by different rules that do not affect these decisions):

1. The Web site must be available 24/7 except for one 12 a.m. to 7 a.m. shutdown each month, for which at least 12 hours notice will be given.
2. Access to the Web site for queries and input must not be interrupted by more than 10 minutes, and the availability for queries and input must be at least 0.9995 when up to 10 users are logged on.
3. When the Web site is unavailable under number 2, a message announcing the time of the expected resumption of service must be displayed within 10 seconds of the onset of the service disruption.

The availability is to be computed quarterly for an expected uptime of 2,150 hours. The quarterly allowable downtime is $0.0005 \times 2,150 = 1.075$ hours = 65 minutes. For the first deviation from these conditions, the customer will be refunded $10,000, for the second one $20,000, and for three or more deviations, the entire quarterly cost will be waived (at a loss of $100,000 to the provider). Each complete 10 minutes of an outage shall be counted as a separate event (e.g., a 25-minute outage will count as two events).

The customer provides the communication lines to the site and failures in these are not covered by the service agreement. The lines inside the site are redundant and their reliability is included in that of the input buffers.

10.1.1 Analysis of the Commercial Power Supply

The server office is located in an area that has frequent power interruptions and therefore an uninterruptible power supply (UPS) has been installed that is capable of sustaining full operation for 2 hours. However, some outages have been longer and have forced suspension of services. Management is concerned about the losses that might arise and has asked you, as head of the reliability department, to make recommendations for upgrading the back-up power facilities.

When you ask the local utility for information on outages, you do not get much cooperation. They report only those that affect more than 30,000 customers and most incidents at your plant are not within that criterion. You find computer logs that help you construct an outage history for the last 2 years, shown by calendar quarters in Table 10.1.

By scanning the last column of the table, you determine that there has been no significant change in the incidence of outages over these 2 years, and thus you deal with the population as a whole. The longest outage duration shown in the table is between 8 and 10 hours, but because of the limited set of data you would like to fit a statistical distribution that might give more insight into the probability of outage durations. The normal distribution does not

Table 10.1
Commercial Power Outages

Quarter	Number of Outages by Duration (Hours)					
	<2	2–4	4–6	6–8	8–10	Total
Current (X)	4	2		1		7
X – 1	3		1	1		5
X – 2	6	1	1			8
X – 3	4	2			1	7
X – 4	5	1		2		8
X – 5	3	3		1		7
X – 6	7	1	1			9
X – 7	3	1	2			6
Total	35	11	5	5	1	57
Average	4.38	1.38	0.63	0.63	0.13	7.13

appear suitable because it permits negative values of the variate, a condition not applicable to outage durations. The exponential distribution, which starts at zero, appears suitable from scanning the bottom row. Only the mean is required to construct this distribution. Using the midpoints of the ranges, you calculate the mean from

$$m = \frac{\sum n_i x_i}{\sum n_i} \tag{10.1}$$

where n_i is the number of outages in each range and x_i is the midpoint of the range. For the numerator of the expression we obtain

$$38 \times 1 + 11 \times 3 + 5 \times 5 + 5 \times 7 + 1 \times 9 = 140,$$

and the denominator is 57, the total shown in the last column. Thus, the mean outage duration $m = 2.46$. The function $e^{-t/m}$ is plotted in Figure 10.1, along with a rendering of the actual outage experience from Table 10.1. Plot points for the experience were obtained by summing the number of outages beyond the endpoint of a range and expressing this as a fraction of the total outages. Thus, 11 outages exceeded 4 hours and the probability of exceeding 4 hours is plotted as $11/57 = 0.19$.

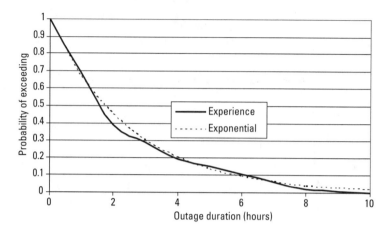

Figure 10.1 Outage duration plot.

It is seen that the exponential distribution provides a workable fit for the actual outage experience. But whereas the experience plot provides no data beyond the end of the 8 to 10 hour range, the exponential distribution indicates that the probability of outages exceeding 10 hours is approximately 0.018.

Your investigation shows that in every one of the eight quarters there have been at least two incidents in which the outage exceeded the 2-hour capacity of the current UPS and therefore you look for ways to tolerate longer outage durations. Essentially two alternatives are available: additional battery capacity for the current UPS and a gasoline-powered generator to augment the current UPS. The estimated quarterly cost and capabilities of these alternatives are listed in Table 10.2. The generator you selected has both automatic and manual start provisions, thus greatly increasing its start-up reliability. For the batteries, the probability of exceeding a given outage duration was obtained from the exponential distribution. For the generator alternative, it is the expected failure probability during operation. The expected cost of outage is the probability of exceeding multiplied by the penalty, computed for two assumptions:

1. Minimum with 50% probability of incurring a penalty of $20,000 and 50% probability of incurring a penalty of $100,000, for an expected value of $60,000.

2. Maximum with 20% probability of incurring a penalty of $20,000 and 80% probability of incurring a penalty of $100,000, for an expected value of $84,000.

The total cost is the sum of the quarterly cost and the expected cost of outage.

Table 10.2
Alternatives for Tolerating Longer Outage Duration

Additions to Current UPS	Quarterly Cost	Probability of Exceeding	Expected Cost of Outage		Total Cost	
			Min.	Max.	Min.	Max.
4-hour battery	$600	0.09	$5,400	$7,560	$6,000	$8,160
6-hour battery	$900	0.04	$2,400	$3,360	$3,300	$4,260
8-hour battery	$1,200	0.018	$1,080	$1,512	$2,280	$2,760
Generator	$1,800	0.01	$600	$840	$2,400	$2,640

Under the minimum total cost assumption, the 8-hour battery addition is the cheapest alternative and under the maximum cost assumption, the generator is preferred. You discuss the findings with management and they select the generator alternative for the following reasons:

- Not all costs of an outage are accounted for by the penalty provisions; there is also loss of revenue and customer good will.
- The generator can cover outages of much more than 10-hours duration, and these cannot be ruled out.

10.1.2 Server Equipment Availability

You have been commended for your analysis of the power outage problem and been given the assignment to evaluate equipment availability. The following are significant differences between equipment availability and power availability:

- Equipment repair is under the company's control;
- Failure and repair data for at least 5 years are available;
- You have many options for improving equipment reliability.

Your starting point is the current installation shown in Figure 10.2, for which you make the following notes:

1. The I/O buffers and servers each represent an active/standby configuration with automatic switching, similar to that shown in Figure 2.4.

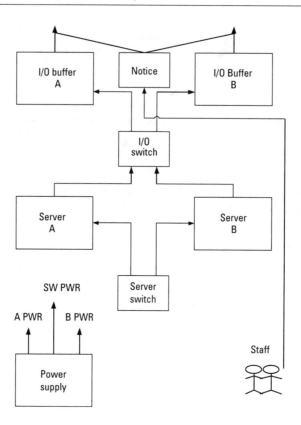

Figure 10.2 Original server equipment installation.

2. The buffers and servers are separately switched, thus providing equipment availability as long as at least one buffer and one server are active.

3. With the changes described in Section 10.1.1, the primary power supply is assumed to be 100% available; the failure probability of the circuit breakers that control the A, B, and SW outputs will be discussed later. Similarly, the reliability of the "Notice" block will be deferred for the time being.

This permits equipment availability to be evaluated in terms of the simple block diagram shown in Figure 10.3. Data for the evaluation of the block availability is shown in Table 10.3.

All of the I/O buffer repairs that required more than 2 hours were completed within 3 hours, but among the switch repairs there were several that required up to 36 hours. For the preliminary availability analysis, you assume that all repair times were at the upper end of the range. The conditions that will make a block (either buffers or servers) unavailable are shown in Table 10.4.

Figure 10.3 Top-level block diagram for equipment availability.

Table 10.3
Availability Parameters

Parameter	I/O Buffer	Server
Item failure rate, hr^{-1}	0.0002	0.0005
Item repair fraction, <0.5 hr	0.2	0.2
0.5 – 1.0 hr	0.3	0.4
1.0 – 1.5 hr	0.3	0.3
1.5 – 2.0 hr	0.1	0.1
>2 hr	0.1	0
Switch failure rate, hr^{-1}	0.0005	0.0005
Switch repair fraction <0.5 hr	0	0
0.5 – 1.0 hr	0.4	0.4
1.0 – 1.5 hr	0.3	0.3
1.5 – 2.0 hr	0.2	0.2
>2 hr	0.1	0.1

Table 10.4
Conditions Leading to Block Failure

Item A fails and		Item B fails and	
Item B is in repair	Switch is in repair	Item A is in repair	Switch is in repair

Because of the symmetry we can compute the probability for one item and then double the result. The cancellations are shown in Table 10.5. The equipment operates 2,150 hours each quarter, and therefore the expected number of failures is as follows: buffer – 0.43, server – 1.08. The average repair time shown in the table is computed by summing the product of the upper limit of the

Table 10.5
Calculation of Downtime

Parameter	Buffer	Server
Expected number of failures (item)	0.43	1.08
Average repair time (item)	1.35	1.15
Failure rate (item)	0.0002	0.0005
Expected downtime (item)	0.00027	0.000575
Average repair time (switch)	4.85	4.85
Failure rate (switch)	0.0005	0.0005
Expected downtime (switch)	0.0024	0.0024
Total expected downtime	0.0053	0.0060

Note: Unit of time is hours.

repair duration and the associated probability. For the buffer the upper limit of the 2-hour interval was 3 hours; for the switch, it was 36 hours.

The total expected downtime is the sum of the shaded rows multiplied by two, as explained in connection with Table 10.4. The total expected equipment downtime is the sum of the two entries in the bottom row, 0.0113 hours per quarter or 0.68 minutes. This does not pose a significant risk of incurring a penalty under the service agreement, and no equipment changes are recommended. However, because the switches contribute the bulk of the downtime, and because the switch downtime is dominated by occasional very long outages, a manual switching capability may be considered. This can be implemented as shown in Figure 10.4.

Now we turn our attention to the circuit breakers in the power supply. For the magnetic circuit breakers that are used, our reliability data book lists an hourly failure rate of 0.2×10^{-6} [1] and this is evenly divided between open failures and failure to operate when required [2]. The open failures will affect only one branch of the installation and will thus by themselves not cause a loss of service. Failure to operate will affect the installation only when there is a short circuit or other overload. No such events have been reported during the past 5 years. The circuit breakers are tested at least once a month and no failure to operate has been noted. Thus, this function does not pose a liability with respect to the service agreement.

Our next concern is the outage notification function. This is currently handled by an interface unit that is a smaller version of the I/O buffer with an attached keyboard. Staff provides free format notices of scheduled maintenance

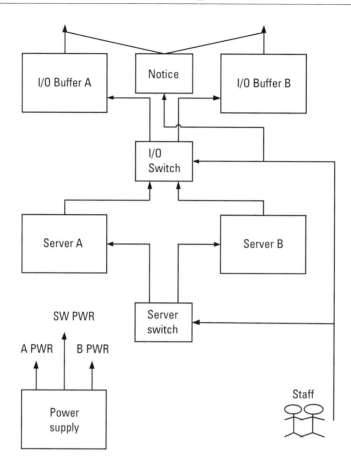

Figure 10.4 Modified server equipment installation.

and expected resumption of service. This implementation was not intended to meet the requirements of notification within 10 seconds of the onset of a malfunction. Two alternatives are considered for this function:

1. Fully automated—activity monitors are permanently attached to each of the two buffers and each of the two server bays. If both buffers or both servers are down, resumption of service is announced for 15 minutes after the start of the outage. This alternative involves purchasing five commercially available activity monitors (four active and one spare) and custom engineering of the combining logic.

2. Semiautomated—when one item is down, an activity monitor is attached to the other item that will announce resumption of service 15 minutes after the monitored item is down. This alternative involves

purchasing two monitors (one active, one spare) and requires no custom logic. The time to attach the monitor is estimated to be 2 minutes, and if the second unit goes down during that interval the notification requirement may not be met.

Cost and capability of the alternatives are compared in Table 10.6. The cost figures and outage probabilities are per quarter. The outage probability for the fully automatic alternative is obtained from an estimated failure rate of the custom logic of 0.001 per hour and the expected outage duration of 0.0113 hours per quarter (see discussion following Table 10.5). The outage probability of the semiautomatic alternative is computed from the failure rate of the server (each item) of 0.0005 hr^{-1} and 2 minute = 0.033 hour connection time. The product, 0.0000165, is then multiplied by 2.8 to account for the failure probabilities of the other server and two buffers. Although the semiautomatic alternative has a much higher outage probability, the cost of outage is insignificant compared to the equipment cost. The semiautomatic implementation is clearly the preferred alternative.

Finally we turn our attention to the overall availability requirement of 0.9995. In the introduction to Section 10.1 this had been translated into a maximum downtime of 65 minutes per quarter. In the discussion of Table 10.5 it is seen that the total expected equipment downtime is 0.68 minutes per quarter. With the improved back-up power supply, the expected outage from that source is negligible. Thus, the overall availability requirement is easily met.

10.2 Increasing Maintenance Effectiveness

A pharmaceutical company converted to a fully automated manufacturing process for its lead product about a year ago. The equipment has functioned satisfactorily but the maintenance costs are high because at least four technicians are necessary to keep it running at all times: an instrument technician to calibrate and repair sensors, an electrician to handle the power distribution and the

Table 10.6
Alternatives for the Notification Function

Alternative	Cost		Outage		Total Cost
	Monitor	Other	Probability	Cost	
Fully automatic	$1,000	$1,500	0.000011	$0.11	$2,500
Semiautomatic	$400	0	0.000046	$0.46	$400

controllers and motors, a machinist to take care of drives and belts, and a pneumatics specialist to service the piping and gas supplies. The process runs 24 hours a day, 7 days a week except for one 12-hour shut-down each month. Most of the maintenance and calibration can be accomplished without disturbing the process but there has been an average of two unscheduled outages a month, each of fairly short duration so that lost production could be made up.

In order to maintain this production record the company employs a lead technician in each of the four specialties and four journeymen who rotate shifts so that at least one of each skill category is at the plant at all times. When not engaged in repair, the machinists and instrument technicians do some preventive maintenance but still have considerable slack time. All technicians are encouraged to study manuals and increase their skills with computer-based training. With wages, benefits, and overhead, the journeyman technicians cost about $100 per hour each. On the other hand, an hour of lost production is estimated to cost $2,500. This has discouraged the company from going to on-demand or call-in staffing for the night and graveyard shifts.

As head of the reliability department you are asked to help reduce the maintenance budget "without hurting the bottom line." Having read Chapter 9, you know exactly what management meant by the quoted phrase: you should not reduce the maintenance cost to the point where the loss in production exceeded the savings in maintenance.

You look at maintenance records for the past year and find the most frequent entries for the off-hour shifts in each of the skill categories, as shown in Table 10.7.

Because the instrument technician and machinist perform preventive maintenance when not engaged in repair, you look first at the duties of the other two categories and divide those into three classes: actions that can be deferred to

Table 10.7
Most Frequent Repair Actions

Instrument Tech.		Electrician		Machinist		Pneumatics Tech.	
Reset scale	48	Replace lamp	102	Feed vibrator	28	Vacuum leak	53
Color scan	30	Small motor	57	Roller bearing	22	Air leak	42
Particle ctr.	20	Medium motor	43	Belt wobble	18	N_2 bottle repl	38
Hygrometer	18	Tank heater	20	Pallet lift	14	Air filter	21
Pyrometer	10	Exhaust fan	8	Positioning	12	Hose repair	10
All others	15	All others	12	All others	10	All others	8

Note: Units are the average number of actions per month.

the day shift, actions that can be assigned to another skill, and actions that require call-in. For the electrician you find that lamp replacement and exhaust fan noise reduction could be deferred. Similarly, for the pneumatics technician the N_2 bottle and air filter replacement could be deferred to the day shift, and at least one-half of the hose repairs could be performed by the day shift if inspection procedures are improved. This rearrangement requires no additional cost and incurs no loss of production.

The remaining electrical and pneumatics activities are shown in Table 10.8 with the recommended disposition.

A total of 10 residual failures were assigned to call-in (six electrical and four pneumatic). The technicians residing closest to the plant are being given a bonus if they respond within 1 hour. With this arrangement, the loss of production is estimated at 10 hours or \$25,000 per month. The savings from eliminating two technicians in each of the two shifts is 4 × 160 × 100 = \$64,000 per

Table 10.8
Disposition of Electrical and Pneumatic Repairs

Repair Item	Assign To	Notes
Electrical		
Small motor	Mechanical	1
Medium motor	Mechanical	1
Tank heater	Instrument	2
All others (50%)	Mechanical	3
All others (50%)	Call-in	
Pneumatic		
Vacuum leak	Instrument	3
Air leak	Instrument	3
Hose repair (50%)	Mechanical	3, 4
All others (50%)	Instrument	3
All others (50%)	Call-in	

Notes:
1. The motors have quick-change drive interfaces, are held in place with cam-locks, and have plug/receptacle electrical connections.
2. The instrument technician is familiar with the tank and usually diagnoses the heater failure. The heater is a screw-in element.
3. Temporary repairs; permanent repairs by day shift.
4. The other 50% of repairs were avoided by inspection.

month. This difference allows for a raise for the instrument technicians and machinists who take on additional duties.

10.3 Replacement of Communication Satellites

The life of communication satellites is limited by three factors:

- Propellant for station keeping;
- Decay of solar cells;
- Wear-out of traveling wave tubes (TWTs).

With regard to each of these, the operator supplies enough reserves so that at least the nominal life is ensured.

In addition, satellites are subject to random failures. Redundancy is the primary means of protection against random failures in vital systems but the degree of redundancy is limited by weight considerations. Usually only dual redundancy is provided, and once one string has failed a replacement launch is scheduled that hopefully will arrive at the station before a second failure of the same function occurs.

The scheduling of replacement satellites had been a hit-or-miss affair, with decisions made under severe time constraints. As head of the reliability department of an operator of communication satellites, you have now been asked to develop contingency plans for launching satellites when either observable wear-out effects or random failures indicate that the satellite may fail soon. The typical event that initiates consideration of a replacement launch is the failure of one element in a redundant subsystem or major component.

The factor that deters immediate procurement of a replacement is the capital investment. On the other hand, it takes considerable time to build and test a satellite and to arrange for a launch. If the active satellite fails before a replacement is available the company has to lease bandwidth on other satellites, and that causes loss of revenue.

You assemble data relevant to the assignment and come up with the following list:

- Satellite design life (exhaustion of expendables) $D = 10$ years;
- Satellite cost, including launch $C = \$25 \times 10^6$;
- Time interval, procurement to launch $T = 1$ year (all critical parts are available);
- Company cost of funds, $i = 10\%$/yr;
- Loss of revenue in case of satellite outage, $L_f = \$10 \times 10^6$/yr;

- Cost of failure prevention (advancing satellite procurement), $V_r = C \times i = \$2.5 \times 10^6/\text{yr}$.

None of these values are to be taken as representative of actual satellite properties for which [3] and [4] may be consulted. They have been selected to facilitate the calculations presented next.

In Chapter 9 you learned that the expenditure for reliability should not exceed the expected cost of failure. You want to find the decision point at which $V_f = V_r$. To the latter we have already assigned a value. To find V_f we make use of (9.1), from which (disregarding the expectation operator E) we have $V_f = f \times L_f$. Since L_f is known, we can write the conditions for the decision point as

$$f = V_r / L_f \qquad (10.2)$$

Using the given data, $f = 2.5/10 = 0.25$ (both numerator and denominator are in units of $\$10^6/\text{yr}$). Thus, only if the failure probability exceeds 0.25 per year is it economically desirable to procure a new satellite.

Next you estimate the single-string failure probabilities of the critical satellite systems, as shown in Table 10.9.

The communication payload is not included in this listing because it is partitioned such that no single failure will cause complete outage and partial failures can be accommodated by using on-board spares. The structures' subsystem is not included because it does not employ redundancy and there are no observable events that change the failure probability. You conclude that there is no single failure in a redundant subsystem that would warrant procurement of a replacement satellite. There are six combinations of dual failures that should lead to procurement of a spare, and all triple failures should.

Table 10.9
Failure Probabilities for Selected Satellite Subsystems

Subsystem	Failure Probability Per Year
Attitude control	0.18
Propulsion	0.14
Electric power	0.12
Thermal control	0.08
Command and control	0.05

When you present your report to management they praise your methodical approach but ask why your recommendation does not depend on the on-orbit time of the satellite. You respond that the potential amount of loss upon complete failure of the satellite does indeed depend on time after launch. A failure during the first year of operation causes a much greater loss than one during the ninth year. But the failures that may cause that loss are random failures, and the probability of a random failure during the next year (the procurement interval T) is the same during the first year as during any year thereafter.

Because of the expected exhaustion of expendables at the end of the tenth year, a replacement satellite will have to be procured at the end of the ninth year. Failures during the tenth year do not require a procurement decision. Procurement decisions during the ninth year are still governed by (10.2) because both the cost of capital and the loss upon satellite failure decrease linearly from the beginning of the ninth year to the beginning of the tenth year.

You state that the value of your replacement strategy is very much dependent on the assumption of random failures (exponential distribution). The available data does not contradict this assumption but neither does it prove it. Therefore you recommend increased instrumentation for localizing failures within each subsystem, and this recommendation is accepted.

10.4 Chapter Summary

This chapter deals with cost trade-off investigations that arise in typical system reliability assignments. In most of these problems there are no unique answers because decisions have to be made on short notice and not all the data required by textbook methods may be available. The emphasis has therefore been on making and explaining reasonable assumptions, keeping to the best-known statistical distributions, and concise representations of results.

In the example in Section 10.1.1 sparse outage data was fitted to the exponential distribution, leading to a reasonable estimation of the probability of outage durations that were within and beyond the range of the observed values. In the second part of Section 10.1 the key to a successful approach was to recognize that outages in a maintained redundant system occur only when the second unit fails while the first one is in repair. Therefore availability could be improved by providing manual switching capability (a very low-cost modification) to overcome the potential threat of a 36-hour outage to replace the automatic switch. Availability can be improved by increasing MTBF or by decreasing MTTF, with the latter usually being easier to achieve and verify by test.

In Section 10.2 the entire problem revolved around maintenance scheduling. The keys to cost reduction were the distinction between maintenance that could be deferred for a few hours and that which required immediate attention;

among the latter, the distinction was between activities that require unique skills and those that could be assigned to an allied skill category.

The final example, procuring replacement satellites, illustrates the benefit of organizing the relevant data for an assignment and also teaches the distinction between incurred cost (that depend on the on-orbit time of a failure) and cost prevention that, for the exponential distribution, is independent of the on-orbit time.

References

[1] Department of Defense, *Military Handbook, Reliability Prediction for Electronic Equipment*, MIL-HDBK-217F, December 1991. This handbook is no longer maintained by the Department of Defense but is still widely used.

[2] Chandler, G., et al., *Failure Mode/Mechanism Distributions*, Rome, NY: Reliability Analysis Center (A DOD Information Analysis Center), 1991.

[3] Wirtz, J. R., and W. J. Larson (eds), *Space Mission Analysis and Design (3rd ed.)*, Boston, MA: Kluwer, 1999.

[4] Wirtz, J. R., and W. J. Larson (eds), *Reducing Space Mission Cost*, Boston, MA: Kluwer, 1996.

11

Applications

The preceding chapters were intended to familiarize the reader with techniques for the successful practice of system reliability engineering and failure prevention. In this last chapter, we draw on these techniques to find solutions to problems that arise in the workplace. A common thread of the examples in this chapter is that there is not a direct correspondence between the stated requirements and the reliability engineer's decision space. Thus some preliminary work is required to really define the problem, giving us a chance to apply more of the knowledge from prior chapters.

In the first example our objective is the reliability design of a power supply for a communications tower. There are availability requirements for the communication services, and from these and details of the other tower components, we establish availability and reliability requirements for the power supply. We examine two alternatives and make a recommendation. Along the way we find that protective devices are needed when power from redundant supplies is combined and we learn how to assess the effect of these devices on system reliability. We also perform a sensitivity analysis to show that the basis for our recommendation is not adversely affected by changes in predicted reliability for a new component.

The second example deals with the electronics bay of an executive jet designed to fly up to 10 hours nonstop. The reliability requirements are stated for the functions of the electronics bay, and we have to translate these to the component level by accounting for the reliability of the power supply. Because of weight and power constraints in the airborne environment, an obvious solution that depends on multiple redundancy is rejected in favor of one that combines redundancy and use of improved parts.

In the last example our objective is the reliability design of a spacecraft attitude determination system. We examine various forms of redundancy and find that an array of six gyros mounted along the axes of a regular tetrahedron has a lower system failure probability than triple redundant gyros mounted along orthogonal axes (a total of nine gyros). In the process we recollect what we learned in earlier chapters about redundancy, and particularly about *k-out-of-n* redundancy.

The chapter summary is intended to highlight the contributions that these examples make to the practice of system engineering and failure prevention.

11.1 Power Supply for Ground Communications

A communication system for an urban area employs transponders mounted on towers and power supplies located in the lower portion of the towers. The system requirements relevant to design for reliability are listed in order of importance:

1. Operational unavailability for 100% standard traffic, $U_{100} < 0.0005$; for at least 50% standard traffic, $U_{50} < 10^{-6}$.
2. Probability of box replacement, $B < 0.5$ per 1,000 hours.

The system reliability engineer has been asked to recommend a power supply configuration that will meet the availability requirements and, if possible, stay under the box replacement allowance. In the following subsections we first discuss the framework within which the power supply selection has to be made, then describe the alternatives, and finally select the configuration.

11.1.1 Framework for Power Supply Selection

To ensure round-the-clock operation, the transponders are triplicated. Each of the three channels is capable of carrying 50% of the standard traffic so that normal operation can be maintained with just two channels operative.

The operation of each channel is monitored centrally over redundant VHF links that are considered 100% reliable. On detection of a failure a repair crew that can restore service within 10 hours ($MTTR = 10$) is dispatched. In addition, a maintenance crew that can check on the condition of equipment visits each tower once a week.

The selection of the transponders has already been made. The MTBF of each channel is 10,000 hours, and from this we compute the steady-state availability for a single transponder in accordance with (2.7) as

$$A_{T1} = MTBF / (MTBF + MTTR) = 10{,}000 / (10{,}000 + 10) = 0.999 \quad (11.1)$$

The transponder function will be available at 100% capacity as long as either (i) all three transponders are available, or (ii) one transponder fails and the other two remain available. Condition (ii) can arise in three ways and therefore that part of the equation must be multiplied by three.[1] Thus,

$$A_{T100} = A_{T1}^{3} + 3 \times A_{T1}^{2} \times (1 - A_{T1}) = 0.997003 + 3 \times 0.000998 = 0.999997$$
$$\text{and } U_{T100} = 1 - A_{T100} = 0.000003 \quad (11.2)$$

The unavailability with regard to the 50% capacity requirement can be computed from the probability that all three transponders become unavailable at the same time. Thus,

$$U_{T50} = U_{T1}^{3} = (1 - A_{T1})^{3} = 1 \times 10^{-9} \quad (11.3)$$

The replacement need for a single transponder during the specified 1,000-hour interval is

$$B_{T1} = t / MTBF = 1{,}000 / 10{,}000 = 0.1 \quad (11.4)$$

For the ensemble of the three transponders this becomes

$$B_{T} = 3 \times B_{T1} = 0.3 \quad (11.5)$$

11.1.2 Power Supply Alternatives

Each of the three transponders requires 60 volt dc (VDC) at 2A and 5 VDC at 6A. These requirements can be met by providing a central supply to service all three transponders, and this will be referred to as the common power supplies. A second alternative is to associate a separate power supply with each transponder, and this alternative will be referred to as dedicated power supplies.

Common Power Supply

For the common power supply alternative it is possible to use commercial power converters that are available from a well-established vendor with capacities of

---

1. See also (6.6a).

10A for the 60-VDC supply and 25A for the 5-VDC supply. The vendor has reliability test data that show that each of the units, when supplying full load, has an MTBF of 40,000 hours. Although a single 60-VDC unit and a single 5-VDC unit have sufficient capacity for all three transponders, they will be employed in a redundant architecture, as shown in Figure 11.1.

Isolating diodes are inserted into the output of each converter to prevent an internal short circuit from pulling down the common bus (shown as a heavy line in the figure). Circuit breakers are provided at each transponder input to prevent short circuits in these components from affecting the bus. The buses are double-insulated to preclude any possibility of a short to ground. With these additions, any single failure—whether in power supplies, transponders, or the protective components—will still provide 100% communication capacity.

The diodes and circuit breakers have failure rates of the order of 10^{-7} per hour that can be neglected in a preliminary analysis. The effect of failures in these components on system availability will be discussed later. The system can then be represented by the RBD shown in Figure 11.2. The transponders have been treated as a single block since the availability had been calculated earlier (11.1–11.3). The 60-VDC and 5-VDC power blocks are identical in both structure and unit MTBF; in the calculations they can be treated as a generic power supply block. Also, since any unit in each supply can provide enough power for all three transponders, the power supply availability can be computed without reference to the 100% and 50% capacity requirements.

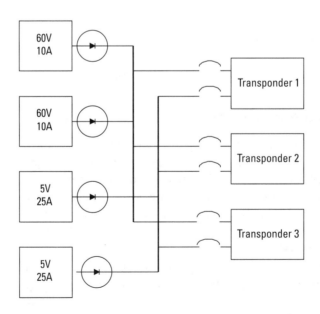

Figure 11.1 Common power supplies.

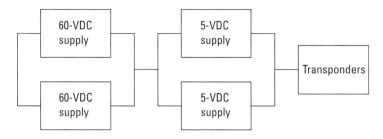

Figure 11.2 RBD for common power supplies.

The availability of a single generic power supply is

$$A_{GP1} = MTBF / (MTBF + MTTR =) = 40,000 / 40,010 = 0.99975 \quad (11.6)$$

The corresponding unavailability $U_{GPI} = 0.00025 = 0.25 \times 10^{-3}$. A generic power supply block (either 60 VDC or 5 VDC) will become unavailable only if both its units become unavailable at the same time. Thus, the unavailability of a generic block is

$$U_{GP} = U_{GPI}^2 = 0.0625 \times 10^{-6} \quad (11.7)$$

Power to the transponders will be lost if either the 60-VDC supply or the 5-VDC supply becomes unavailable. Thus,

$$U_P = 2 \times U_{GP} = 0.125 \times 10^{-6} \quad (11.8)$$

As already mentioned, this unavailability is total; there is no distinction between the 100% capacity and 50% capacity states. However, at the system level, when the power supply availability is added to the transponder availability, the distinction re-emerges. Thus,

$$U_{100} = U_P + U_{T100} = 0.125 \times 10^{-6} + 3 \times 10^{-6} = 3.125 \times 10^{-6} \quad (11.9)$$

$$U_{50} = U_P + U_{T50} = 0.125 \times 10^{-6} + 1 \times 10^{-9} = 0.126 \times 10^{-6} \quad (11.10)$$

The box replacement probability over a 1,000-hour interval for the four power supply components, each with MTBF of 40,000 hours is [see (11.4) and (11.5)]

$$B_P = 4 \times 1,000 / MTBF_{PG} = 4 \times 0.025 = 0.1 \quad (11.11)$$

The total system box replacement probability is then computed as

$$B = B_P + B_T = 0.1 + 0.3 = 0.4 \qquad (11.12)$$

We return now to the failure effects of the protective components shown in Figure 11.1, the diodes and circuit breakers. Power diodes have essentially two failure modes: open and high reverse current (leakage). Circuit breakers also have two predominant failure modes: open and failure to operate on overload (fail closed). The effects of the failures in these protective devices are analyzed in an FMEA worksheet shown as Table 11.1. The format of this worksheet is modified from that described in Chapter 5 in the following ways:

- Because the system is very simple, only failure effects at the system level are described.
- The compensating provisions column has been eliminated because the low loss of service probability (see following text) does not warrant redundancy for the protective components.

Since this application does not involve human safety, severity I has been assigned for complete loss of service, severity II for partial loss of service (at least 50% capacity is available), severity III for service interruptions of less than 2 minutes, severity IV for other service impairments, and severity V for no service impairment.

The diode open failure and circuit breaker open modes have been classified as severity IV (negligible) because there is an impairment of the capability to handle higher-than-standard traffic (in case of the circuit breaker) and there may

Table 11.1
FMEA of Protective Components

ID	Part	Failure Mode	Effect	Severity	Detection	Remarks
1.1.1	Diode	Open	No voltage	IV	Monitor	
1.1.2	Diode	Leakage	None	V	Inspection	Severity I if supply output shorted
1.2.1	Circuit breaker	Open	No voltage	IV	Monitor	
1.2.2	Circuit breaker	Fail closed	None	V	Inspection	Severity I if transponder is shorted

be a reduction of voltage, leading to increased error rates (in case of the diode). Detection by monitor implies a continuous surveillance of equipment status. For the open diode this can be accomplished by sensing current flow in each of the power supply outputs. For the open circuit breaker it is accomplished by monitoring the output of each of the transponders. Detection by inspection makes use of the weekly visits by a maintenance crew. For the diodes it requires removal, one at a time, from the circuit and measuring the reverse current under specified voltage conditions. The sensitivity of this detection method can be improved if data are collected to detect trends of increasing reverse current. Snap-in mounts for diodes are available to facilitate the removal. Inspection of the circuit breakers is accomplished by use of push-to-test provisions.

We will now estimate the loss of service probabilities associated with failures in the protective components. Open failures (ID 1.1.1 and 1.2.1) are detected as soon as they happen and are repaired within 10 hours. Loss of service due to a diode open failure will occur only if the other power supply for that voltage experiences a failure during the repair interval. The probability of this failure is

$$P_{f1} = t_R / MTBF_p = 10 / 40{,}000 = 0.25 \times 10^{-3} \tag{11.13}$$

where t_R is the repair time.

A conservatively rated power diode has a failure rate λ_D well below 10^{-7} per hour [1] and we will use that number as a upper limit. We also assume that one-half of the diode failures result in an open condition and the other half will result in a high reverse current condition. These are very conservative assumptions because many diode failures result in degraded parameters rather than the extreme conditions postulated here. Using P_{Do} for the probability of diode open failure during 1 year (8,760 hours) we obtain

$$P_{Do} = 8{,}760 \times \lambda_{Do} = 8{,}760 \times 0.05 \times 10^{-6} = 0.44 \times 10^{-3} \tag{11.14}$$

This quantity represents the probability that a repair for an open diode will occur during the year. The 1-year loss of service probability due to an open diode is therefore

$$P_{XDo} = P_{f1} \times P_{Do} = 0.25 \times 0.44 \times 10^{-6} = 0.11 \times 10^{-6} \tag{11.15}$$

Service can be restored in 10 hours, and the unavailability due to this loss is therefore

$$U_{Do} = t_R \times P_{XDo} / 8{,}760 = 1.1 \times 10^{-6} / 8{,}760 = 0.13 \times 10^{-9} \tag{11.15a}$$

An open circuit breaker will cause capacity to decrease to 50% if one of the other two transponders fails during the 10-hour repair interval. Similar to (11.13), we have

$$P_{f2} = 2 \times t_R / MTBF_T = 20 / 10{,}000 = 2 \times 10^{-3} \qquad (11.16)$$

The failure rate for conservatively applied magnetic circuit breakers in a ground environment is $\lambda_C = 0.02 \times 10^{-6}$ per hour [1]. Again, we assign one-half of this to open failures and the other half to failure-to-open. Thus, the probability of an open circuit breaker failure during 1 year

$$P_{Co} = 8{,}760 \times \lambda_{Co} = 8{,}760 \times 0.01 \times 10^{-6} = 0.088 \times 10^{-3} \qquad (11.17)$$

The probability over 1 year of not having full capacity due to an open circuit breaker failure is therefore

$$P'_{XCo} = P_{f2} \times P_{Co} = 2 \times 10^{-3} \times 0.088 \times 10^{-6} = 0.176 \times 10^{-9} \qquad (11.18)$$

The corresponding unavailability, calculated as in (11.15a), $U'_{Co} = 0.02 \times 10^{-12}$. The prime on the P'_{XCo} and U'_{Co} symbols signifies that this outage probability is for 100% capability. An even lower outage probability applies to the 50% capacity requirement.

The calculations shown in (11.13–11.18) demonstrate that it was indeed justified to neglect the failure probability of the protective devices in the availability calculations. This is similar to the conclusions we arrived at in Section 10.1.2. Where the state of protective devices can be monitored or inspected at intervals that make the existence of a latent failure extremely unlikely, their availability is generally so much higher than that of the monitored equipment that the presence of these devices can be neglected in the calculations. However, where protective devices are used in safety-critical applications, additional safeguards may be required.

Dedicated Power Supplies

For the dedicated alternative, the manufacturer of the transponder has identified a supplier for a custom design that will furnish both voltages with about 25% reserve power capacity in a single unit. There are no reliability test data available, and the supplier claims a parts count MTBF prediction of 35,000 hours. Because supplier predictions tend to be on the optimistic side we will use a "working hypothesis" of 25,000 hours. Each of the power supply units will be directly connected to a transponder, yielding the dedicated power supply configuration shown in Figure 11.3.

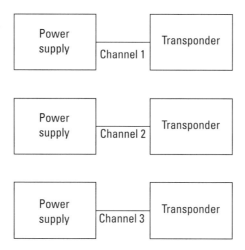

Figure 11.3 Dedicated power supplies.

An attractive feature of this architecture is that the channels are physically and electrically independent. But with regard to availability this arrangement introduces undesirable coupling because during a replacement of a power supply any failure in one of the other channels will reduce the capacity below the 100% level. The following analysis will further clarify this problem. The architecture shown in Figure 11.3 must be analyzed in terms of channel availability rather than component availability. We start by computing the channel MTBF

$$1/MTBF_C = 1/MTBF_T + 1/MTBF_P = (0.1 + 0.04) \times 10^{-3} = 0.14 \times 10^{-3}$$
(11.19)

From which $MTBF_C = 7{,}140$ hours. Again making use of (2.7), we obtain the availability of a single channel as $A_C = 0.9986$.

The channel unavailability is computed as $U_C = 1 - A_C = 0.0014$.

The 100% capacity can be maintained if either of the following conditions is met: all three channels are available or one channel is unavailable and the other two are available. The latter condition can arise under three circumstances (channels 1, 2, or 3 unavailable) Thus,

$$A_{100} = A_C^3 + 3 \times U_C \times A_C^2 = 0.995806 + 0.004188 = 0.999994$$
(11.20)

From which

$$U_{100} = 1 - A_{100} = 6 \times 10^{-6}$$
(11.21)

At least 50% capacity will be maintained as long as at least one channel remains operative. It is convenient to compute the 50% unavailability as the probability of all three channels becoming unavailable. Thus

$$U_{50} = U_C^3 = 2.7 \times 10^{-9} \tag{11.22}$$

The box replacement need for the power supplies is a function only of their number and MTBF; it is not affected by the configuration in which the boxes are used. Thus, applying (11.4) and (11.5) to the power supply MTBF of 25,000 hours we get

$$B_{P1} = 1{,}000 \,/\, 20{,}000 = 0.04 \text{ and} \tag{11.23}$$

$$B_P = 3 \times B_{P1} = 0.12 \tag{11.24}$$

When added to the transponder box replacement probability (11.5) the total replacement probability for a 1,000-hour interval then becomes

$$B = B_P + B_T = 0.12 + 0.3 = 0.42 \tag{11.25}$$

11.1.3 Evaluation of Alternatives

The principal characteristics of each of the alternatives are summarized in Table 11.2.

It is seen that both alternatives meet the requirements. The common power supply architecture achieves lower unavailability with respect to the 100% capacity requirement, and because this has been identified as the most important characteristic, this alternative is tentatively selected. We continue to evaluate the differences in the other rows. The dedicated supply offers a much lower unavailability with respect to the 50% capacity requirement, but when translated into operating time the difference becomes irrelevant: the unavailability for the common supply of 0.126×10^{-6} is equivalent to less than 4 seconds per year. There is little difference between the alternatives in the probability of box replacement. The common supply requires periodic inspections but these are allowed in the statement of the problem. If the application involved an inaccessible site, the dedicated supply would probably have been preferred. Examination of (11.13–11.18) shows that switching from weekly inspections to monthly ones will have little impact on the protective devices' contribution to overall unavailability.

Although there is not a stated requirement, the design maturity of the commercial converters will frequently be a significant factor in selecting alternatives that are otherwise equally desirable. Design maturity reduces the risk with

Table 11.2
Characteristics of the Power Supply Alternatives

Characteristic	Requirement	Common Supply	Dedicated Supplies
Unavailability – 100% capacity, U_{100}	$<10^{-4}$	3.125×10^{-6}	6×10^{-6}
Unavailability – 50% capacity, U_{50}	$<10^{-6}$	0.126×10^{-6}	0.003×10^{-6}
Box replacement 1,000 hrs, B	<0.5	0.4	0.42
Inspection required	Allowed	Yes	No
Design maturity	No requirement	Yes	No

respect to reliability, as well as performance and delivery. The absence of test data for the dedicated supply was the reason that the parts count prediction of 35,000 hours MTBF was reduced to 25,000 in our calculations.

Before finalizing the selection of the common power supply, we will now repeat the U_{100} calculation for an assumed 35,000-hour MTBF for each dedicated supply. Values for the 25,000-hour and 35,000-hour assumptions are compared in Table 11.3.

As expected, the unavailability is reduced when the MTBF is increased, but even the reduced unavailability for the dedicated supply is significantly higher than that for the common supply. It is concluded that the selection of the common supply alternative is robust with respect to the assumptions about the

Table 11.3
Effect of MTBF Assumptions on Unavailability

Parameter and Equation	MTBF 25,000	MTBF 35,000
$MTBF_C$, 11-19	7,140	7,780
A_C, 11-20	0.9986	0.9987
U_C	0.0014	0.0013
A_{100}, 11-21	0.999994	0.999995
U_{100}, 11-22	6×10^{-6}	5×10^{-6}

MTBF of the dedicated supply. The comparison shown in Table 11.3 is usually called a sensitivity study or sensitivity analysis. It is a highly recommended step in any reliability analysis involving assumed parameters.

11.2 Reliability of Aircraft Electronics Bay

The electronics bay of a long-range corporate jet contains flight instruments and control, communications, and cabin entertainment components. Three functions are essential for safety of flight: flight control, instruments, and emergency communications. It is required that these three functions together have a failure probability of not more than 0.5×10^{-6} for a 10-hour flight segment. The operation of these functions is dependent on the availability of electric power and therefore the reliability of the power supply must be evaluated to arrive at the allowance for the essential electronics bay components. The primary aircraft power supply is 125 volt ac (VAC), variable frequency, furnished by four alternators that are attached, two on each side, to the turbine section of the two jet engines. The primary power is fed to transformer-rectifier units (TRU) that deliver 28 VDC to the electronics bay. Each TRU is rated at 50-amperes steady state, and a single TRU can supply sufficient power for all essential loads.

11.2.1 Primary Power Supply

The design for supplying primary power is shown in Figure 11.4. The relevant random failure rates are shown in Table 11.4. The turbine failure rates are

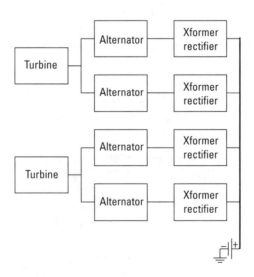

Figure 11.4 Aircraft power distribution.

Table 11.4
Power Supply Failure Probabilities

Component	Symbol	Failures/ 10^6 Flt-Hrs
Turbine	λ_T	20
Alternator	λ_A	1200
Xformer-Rectifier	λ_X	210

predicated on inspection prior to each long-range flight and on compliance with all recommended maintenance procedures. Although a battery is shown in Figure 11.4, its contribution to the power supply cannot be counted on for added reliability because of the limited time it can sustain the loads.

The 10-hour failure probability for one side of the power supply is

$$\lambda_{P1} = 10\lambda_T + (10\lambda_A + 10\lambda_X)^2 = 400 \times 10^{-6} \qquad (11.26)$$

A failure of the entire primary power supply will occur only if both sides fail. Thus,

$$\lambda_P = \lambda_{P1}{}^2 = 160{,}000 \times 10^{-12} = 0.16 \times 10^{-6} \qquad (11.27)$$

11.2.2 Safety Critical Loads

To meet the safety critical function failure probability requirement of not more than 0.5×10^{-6} per 10-hour segment, the essential functions by themselves should have a failure probability of less than 0.34×10^{-6} per segment. Failure rates for a single channel of the safety critical loads are shown in Table 11.5. We will now examine how the reliability requirements can be met.

The 10-hour failure probability of a single-series string of flight controls, instruments, and emergency communications is

$$F_F + F_I + F_C = 0.000980 \qquad (11.28)$$

This falls far short of the requirement. Thus, redundancy must be considered. Redundancy for this configuration requires a modification of our modeling of series strings. Although all three functions are essential for the safety of the aircraft, and are therefore considered in series in (11.28), they operate

Table 11.5
Safety Critical Loads—Single Channel Data

Component	Failure Rate, 10^{-6}		10-Hour Failure Probability	
	Symbol	Value	Symbol	Value
Flight controls	λ_F	50	F_F	0.000500
Instruments	λ_I	40	F_I	0.000400
Emergency communications	λ_C	8	F_C	0.000080

independently (e.g., the use of emergency communications requires services from neither flight controls nor instruments). Thus, redundancy provisions can be introduced independently. The partitioned redundancy shown in Figure 11.5 illustrates one possible implementation, but not every function needs to have the same level of redundancy.

The resulting 10-hour failure probability for the configuration shown in this figure is

$$F_F^2 + F_I^2 + F_C^2 = (0.25 + 0.16 + 0.006) \times 10^{-6} = 0.416 \times 10^{-6} \quad (11.29)$$

and still falls short of the requirement. In examining the individual terms of this expression, it is seen that the dual redundant instruments and emergency communications have a failure probability below the goal of 0.34×10^{-6} for the 10-hour segment. With triple redundancy for the flight controls, the segment failure probability for the safety critical functions becomes

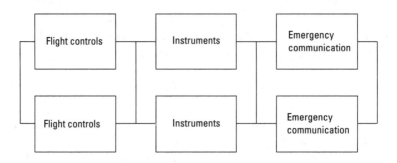

Figure 11.5 Partitioned safety critical loads.

$$F_F^3 + F_I^2 + F_C^2 = (0.000125 + 0.160 + 0.006) \times 10^{-6} = 0.166 \times 10^{-6}$$
(11.30)

This is comfortably below the requirement. But there is a high penalty in terms of equipment cost, weight, and power consumption for triple redundancy of the flight controls. This suggests investigating options that require less equipment, such as use of derating (using parts that have a higher rating and are thus used at only a fraction of their capability) or partial redundancy within a single channel.

Before investigating these possibilities in detail, we want to calculate the maximum 10-hour failure probability that can be tolerated. Since the contribution of the triple redundant flight controls in (11.30) was negligible, we can obtain the maximum allowable failure probability from $(0.34-0.166) \times 10^{-6} \approx 0.17 \times 10^{-6}$. This is the failure probability allowance for redundant flight controls that compares to the current failure probability of 0.25×10^{-6}. The single-string segment failure probability corresponding to the 0.17×10^{-6} is 0.412×10^{-3} or a failure rate of 41.2×10^{-6} per flight hour. This compares with the 50×10^{-6} for the original design; the required reduction of less than 20% appears feasible.

11.2.3 Partial Improvement of a Function

A first-level breakout of the contributors to the flight control failure rate is shown in Table 11.6. The failure rates are expressed in 10^{-6} per flight hour.

The power transistors are currently used at 70% of their rated power dissipation. A higher dissipation type is available that can carry the same load at 25%

Table 11.6
Contributors to Flight Control Electronics Failure Rate

Function	Failure Rate 10^{-6}/hr
Power transistors	16
Signal conditioning	11
Timing and synchronization	10
Channel processing	6
Voltage regulator	6
Miscellaneous	1
Total	50

of its rating, which will reduce the failure rate to 10×10^{-6}. The signal conditioning is a physically small unit but has a high gate count that accounts for the relatively high failure rate. The unit could be made redundant but this will involve considerable switching for managing the redundancy.

Detailed review of the timing and synchronization (T&S) components shows two leading contributors to the failure rate: a phase-locked loop (5×10^{-6}) and a clipping circuit (3×10^{-6}). Circuit improvements and use of newer parts can bring the failure rate of the phase-locked loop down to 2×10^{-6}. The clipping circuit can be replaced by a clipping diode with a failure rate of 0.1×10^{-6}. Thus, the failure rate of the T&S components can be reduced to 4.1×10^{-6}.

At this point the failure rate has been reduced by 6×10^{-6} in the power transistors and by 5.9×10^{-6} in the T&S circuit; this is more than enough to meet the requirements. However, it may pay to continue down the list to look for additional improvement from which the least expensive and risky ones can be selected.

The failure rate of channel processing is dominated by a microprocessor that has been selected for high reliability and for which no better alternatives could be found. However, the data sheet indicates that the failure rate of the function can be reduced to 4×10^{-6} if the temperature of the mounting surface can be reduced by 6C. Improvements in the thermal conduction path to the enclosure make this temperature reduction possible.

In the voltage regulator it is possible to use newer parts that will bring the failure rate of this component down to 4.5×10^{-6}. The result of these improvements is shown in Table 11.7.

The changes in the T&S function were not selected because they involved major revisions of the circuit in addition to use of new parts. Thus, they carried

Table 11.7
Improved Flight Control Electronics Failure Rate

Function	Failure Rate, 10^{-6}		
	Original	Possible	Selected
Power transistors	16.0	10.0	10.0
Signal conditioning	11.0	11.0	11.0
T&S	10.0	4.1	10.0
Channel processing	6.0	4.0	4.0
Voltage regulator	6.0	4.5	4.5
Miscellaneous	1.0	1.0	1.0
Total	50.0	34.6	40.5

a higher risk than the changes in the channel processing and voltage regulator functions that did not involve circuit changes. The resulting single channel failure rate for the flight controls is 40.5×10^{-6} per flight hour, yielding a segment failure probability of 0.405×10^{-3}, which is below the requirement of 0.412×10^{-3}.

11.3 Spacecraft Attitude Determination

Spacecraft reliability is a highly constrained—some say overconstrained—field of endeavor. The following are its distinguishing difficulties:

- Very high reliability requirements because of the high cost of satellite replacement;
- Long life requirements, again related to the high cost of replacement;
- Inability to repair on-orbit (except for space shuttle missions to the Hubble Space Telescope and a few retrievals of failed satellites, all very costly);
- Extreme weight and power constraints;
- Limited reliability experience for contemporary components.

Out of these challenges arose notable advances in reliability engineering, such as launch readiness reviews, Class S parts,[2] and specialized redundancy practices, some of which are discussed next.

Earth observation requires that the instantaneous position of a satellite and the pointing direction of the instruments be known. Both of these measurements require knowledge of spacecraft attitude relative to a geocentric coordinate system. Laser gyroscopes are commonly used for attitude sensing and, as employed in spacecraft, have a failure rate of approximately 5×10^{-6} per hour.

11.3.1 Orthogonal Gyro Configurations

Gyros are conventionally employed in orthogonal configurations, as shown in Figure 11.6. The failure rate of one such triad is 15×10^{-6} per hour. The spacecraft in which they are employed have nominal lifetimes of at least 7 years or 61,320 hours. The failure probability of a single triad is therefore

---

2. Space-qualified parts produced by a closely monitored process and under strict configuration control. The benefits of these practices were so notable that they have been adopted outside the space environment, and therefore the need for Class S parts has become less urgent.

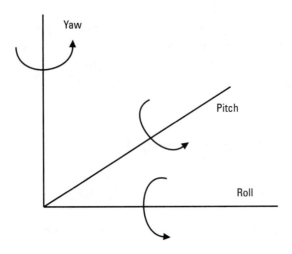

Figure 11.6 Orthogonal gyro reference axes.

$$F_t = 1 - e^{-\lambda t} = 1 - e^{-0.92} = 0.6 \qquad (11.31)$$

Note that the approximation for small failure probabilities (2.2a) does not hold in this case.

A redundancy structure composed of two such triads will still have an unacceptably high failure probability.

$$F_{t2} = F_t^2 = 0.6^2 = 0.36 \qquad (11.32)$$

An alternative approach is to make each gyro redundant such that there are two yaw gyros, two pitch gyros, and two roll gyros. The failure probability for a single gyro will be

$$F_G = 1 - e^{-\lambda t} = 1 - e^{-0.32} = 0.274 \qquad (11.33)$$

And for a redundant pair of gyros it will be

$$F_{G2} = P_G^2 = 0.274^2 = 0.075 \qquad (11.34)$$

The failure probability of the triad of redundant gyros will be three times this value or 0.225. This is an improvement over the 0.36 failure probability for the redundant triads but it is still unacceptable, particularly if missions longer than 7 years are anticipated.

11.3.2 Nonorthogonal Gyro Orientation

If six gyros are oriented along the edges of a regular tetrahedron, as shown in Figure 11.7, the satellite orientation can be determined by coordinate transformations as long as any three of the initial six gyros remain operative.

The attitude determination function will be lost if more than three gyros fail. This is a case of *k-out-of-n* redundancy and the probability can be computed by (6.6). To gain better insight into the operation of this type of redundancy we will assess the probability by separately evaluating the case of six, five, and four gyros failing. The summary of the individual calculations is shown in Table 11.8.

For the case of six gyros failing, the number of combinations is 1 (there is only one way in which all six gyros can fail). The probability of this happening is 0.274^6, which is recorded in the "Failing" column. In this case no gyros remain operational (not failing) and the probability of this is 1. In the table's last column we record the product of the middle three columns and, in this case, it is just one significant term.

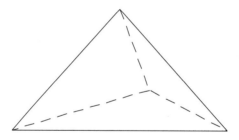

Figure 11.7 Nonorthogonal gyro reference axes.

Table 11.8
Failure Probability for Nonorthogonal Gyros

Number of Gyros Failing	Number of Combinations	Probability of:		Product
		Failing	Not failing	
6	1	0.000423	1	0.000423
5	6	0.001544	0.726	0.006727
4	15	0.005636	0.527	0.044562
Sum of the above				0.051712

When five gyros are failing one gyro remains operational, and since this can be any one of the six original gyros, we deal with six combinations. All other entries in this row are computed similar to those of the preceding one. When four gyros are failing two gyros remain operational, and the applicable number of combinations is 15,[3] and again the other entries in the row are similar to the preceding ones. When three gyros are failing the other three provide enough information for attitude determination and thus no further table entries are required. The row of four failing gyros is clearly the governing term.

Table 11.9 summarizes the results of the three configurations that we have examined in this section. All employ six laser gyros and thus use essentially the same resources. As is acknowledged in the table's last column, the alternatives that achieve the low hardware failure rate have a greater dependence on software. But software failures can usually be overcome by work-arounds or uploading of corrections or patches, while hardware failures (when not tolerated by means of redundancy or other back-up provisions) usually terminate the useful life of the satellite.

One disadvantage of the nonorthogonal gyros is that accuracy degrades as gyros fail, and when only three gyros remain the degradation can be significant. But in view of the much lower failure probability of the nonorthogonal gyros, this configuration should be high on the list of candidates. Other considerations may also come into play: wider experience with orthogonal gyro configurations, availability of mounting arrangements and attitude determination software, and the absence of accuracy degradation after gyro failures. As a point of reference: a triad of triple-redundant gyros has a failure probability of 0.06 for the 7-year mission, roughly comparable to that of the nonorthogonal gyros but, of course, requiring 50% more components.

Table 11.9
Attitude Determination Failure Probability (7-Year Mission)

Configuration	Failure Probability	Software Requirements
Dual triads	0.360	Minimal
Triad of dual gyros	0.225	Modest
Nonorthogonal gyros	0.052	High

---

3. The number of combinations is the first term after the summation symbol in (6.6). It can also be found under the "Math-Trig" functions in a spreadsheet, such as Microsoft Excel.

11.4 Chapter Summary

In this chapter we applied techniques introduced in earlier chapters in the context of problems that must be faced by the practicing reliability engineer. These problems typically have multiple solutions, no one of which may be clearly the best in an absolute sense. We tried to point to nonquantifiable factors, such as design maturity and prior usage, which sometimes guide the decision away from preferences based on quantifiable factors. Because the ultimate decision is made at the system level, we concluded each example with a summary of the alternatives examined.

Another common thread that runs through all of the examples is that the composition, and hence the reliability, of some elements is fixed, and the achievement of the availability or reliability goal depends on interfacing the "to be designed" portions with the fixed portions.

In the first subchapter, we dealt with a commercial communications tower. The equipment is ground-based (with few restrictions on size and power) and available for repair and inspection. The main emphasis is on continuous availability of the service, the source of the revenue stream.

The second section discusses the reliability of the avionics bay in an executive aircraft designed for flight segments of up to 10 hours. This being an airborne environment, weight and power are at a premium. We deal with this by minimizing the amount of redundancy. Whereas in the previous example the power using equipment was predefined and the source of the electricity was to be optimized, we are faced here with a fixed source of power and have to arrange the using equipment to meet the reliability goals. We found that combining redundancy with component improvement leads to an acceptable solution.

In the third example we dealt with a spacecraft attitude determination system in an environment that puts an even higher premium on weight and power than the aircraft one. We examined several methods of redundancy to meet requirements for a 7-year lifetime for an earth observation satellite. An array of six nonorthogonally mounted gyros offered much higher reliability than any orthogonal dual redundant configuration (also involving six gyros).

All of the examples, and particularly the last one, encouraged thinking "outside the box." That, along with team spirit and collaboration with other disciplines, is what makes a good systems reliability engineer.

Reference

[1] Department of Defense, *Military Handbook, Reliability Prediction of Electronic Equipment, MIL-HDBK-217F,* December 1991. This handbook is no longer maintained by the Department of Defense but is still widely used.

About the Author

Herbert Hecht is vice chairman of the Board of SoHaR Incorporated, an R&D and consulting company for high dependability systems. (The name of the company is a contraction of software and hardware reliability.) In addition to his managerial duties, he supervises technical work in hardware and software reliability analysis, sneak circuit analysis, and safety analysis.

His expertise has caused him to be appointed to many governmental and academic review bodies for safety and reliability, such as the National Research Council's Committee on Engineering Challenges to the Long-Term Operation of the International Space Station (a congressionally mandated committee), 1998–2000, and the Nuclear Regulatory Commission's Expert Panel on Digital System Research, 1999. He has participated in the NASA Ames Workshop for Design for Safety and in safety investigations at the NASA Jet Propulsion Laboratory.

Prior to founding SoHaR, Dr. Hecht was employed at The Aerospace Corporation, El Segundo, California (1962–1977), leaving as director of Digital Systems for Advanced Programs. One of his assignments was the development of the Ascent Guidance Monitoring System for the Gemini Launch Vehicle (GLV), and in that role he participated in the first two launches of the GLV. Earlier, he was employed at the Sperry Corporation as department head for Helicopter Flight Controls. He developed and managed the development of navigation and flight control systems that required extremely high dependability.

Dr. Hecht's professional activities include a term on the Board of Governors of the IEEE Computer Society, service as Visitor in Computer Engineering for ABET, and membership in standards working groups of the IEEE, the AIAA, and the ISA. He has more than 100 papers in refereed journals or conference proceedings and holds 12 patents in the field of highly reliable control systems. He is licensed as a professional engineer in control systems engineering by the state of California.

Dr. Hecht received a bachelor of electrical engineering degree from City College, New York, and a master's degree in the same subject from Polytechnic Institute of Brooklyn (now part of Polytechnic University). In 1967 he received a Ph.D. in engineering from the University of California at Los Angeles, with a dissertation on economics of reliability for space launch vehicles.

Index

Action items, 155
Activity blocks, 138
Adaptive changes, 147
Aircraft electronics bay, 210–15
    channel processing failure rate, 214
    failure rate contributors, 213
    partial improvement of function, 213–15
    power distribution, 210
    power supply failure probabilities, 211
    primary power supply, 210–11
    safety critical loads, 211–13
    timing and synchronization (T&S) components, 214
    See also Applications
Alternates
    active, 97–98
    diverse, 95–96
    dormant, 97–98
    identical, 95–96
The American National Standard Recommended Practice for Software Reliability, 112
Analytical approaches, 37–61
    classes, 37
    failure modes and, 38–52
    fault tree analysis (FTA), 56–61
    sneak circuit analysis (SCA), 52–56
    summary, 61
Analytical redundancy, 105–6
    advantages, 106

    defined, 105
    example, 105–6
    uses, 105–6
Applications, 199–219
    aircraft electronics bay, 210–15
    overview, 199–200
    power supply, 200–210
    spacecraft attitude determination, 215–18
    summary, 219
Audits, 153
Availability
    calculations, 18
    of essential services, 16
    generic power supply, 203
    of repairable system, 15
    requirements, 199
    server equipment, 187–92
Aviation accidents, 29–31
    American Airlines flight 1420, 29
    Hageland Aviation, 30–31
    TWA flight 800, 30
    See also Failure(s)

Base failure rates, 8
Behaviors, 127
Breadboard test, 73
Built-in test (BIT), 55, 83–84
    characteristics, 83
    equipment, 83
    See also In-service testing

Changes
adaptive, 147
perfective, 147
system reliability effects, 147–48
Chernobyl, 27–29
accident cause, 28
defined, 27
operator actions, 28
postaccident safety measures, 28–29
reactor features, 27
See also Failure(s)
Common power supplies, 201–6
availability, 203
defined, 201–2
illustrated, 202
isolating diodes, 202
RBD for, 203
See also Power supplies
Communication satellites
communication payload, 196
failure probabilities, 196
life factors, 195
replacement of, 195–97
replacement scheduling, 195
Component prototype test, 74
Computer-based tools, 46
Concept phase, 137
answers, 142
reliability issues, 141–42
start of, 141
See also Life cycle
Condition tables, 121, 122
Configuration(s)
control, 158
management, 141
orthogonal gyros, 215–16
quadruple voting, 102
Control charts, 159–60
defined, 159–60
use of, 160
in visualization, 160
See also Quality assurance
Cost
elements, estimation, 174–77
expected, of failure, 176
of failure, 167–82
fault tolerance, 131
generic, of reliability model, 177–81
increment for reliability, 168

relations for reliability improvement, 169
testing, 63, 152
total, 186, 187
total user curve, 170
trade-offs, 183–98
Could not duplicate (CND), 72, 145
Coverage
accounting for, 94–95
defined, 94
imperfect, 94–95
imperfect, effect of, 95

Dedicated power supplies, 206–8
box replacement, 208
channel independence, 207
channel unavailability, 207
defined, 206
illustrated, 207
MTBF of, 209–10
See also Power supplies
Design and evaluation tasks, 150
Design margins, 66–70
increased, 174–75, 181
normalized, 67
probability of failure vs., 67
Detector threshold, 68
distributions, 69
mean, 68
Development phase
end of, 144
failure prevention and, 142
reliability and, 137
reliability issues in, 142–44
requirements release, 141
spiral model for, 140
transition to, 141
waterfall model for, 139
See also Life cycle
Discrete parameter model, 14–15
availability computed from, 15
defined, 14
probability calculation, 15
Discrimination ratio, 65
Dual redundancy, 91–98
active vs. dormant alternates, 97–98
dynamic, 91–95
identical vs. diverse alternates, 95–96
static, 91–95
See also Redundancy
Dynamic redundancy, 91–95

defined, 91
illustrated, 92
requirement, 92–93
software application, 129, 131
See also Redundancy

Electronic parts tolerance analysis, 151
Embedded software, 109
End-of-period uniform payments, 173
Engineering change proposal (ECP), 139
Environmental stress screening (ESS), 152
Error detecting and correcting code (EDAC), 82
Error detecting codes (EDC), 82–83
    arithmetic operations for, 105
    characteristics, 83
    in digital operations, 93
    using, 82
Error detection, 93–94
Expenditures
    administrative obstacles and, 171
    benefits and, 171, 172, 173, 174
    financial modeling, 173
    periodic uniform, 173
    single, 171
    time considerations, 170–74
    uncertain benefits and, 171
    See also Cost
Exponential distribution, 5–8
    defined, 6
    illustrated, 6
Extreme programming, 141

Failure analysis and corrective action procedure (FRACAS), 64
    good, 70
    key entries, 70–72
Failure modes, effects and criticality analysis (FMECA), 38
Failure modes and effects analysis (FMEA), 38–52
    alternative approaches, 46–49
    applications, 38
    concepts, 38–39
    documentation, 39
    documents generated from, 50
    for electronic artillery fuze, 48
    failure probability estimates, 42
    flight phase failure effects, 45
    flight phase failure effects summary, 44
    functional approach, 47, 48
    model-based development and, 49
    parts approach, 48
    as plan for action, 49–52
    in process industry, 38
    for project support activities, 49–50
    of protective components, 204
    report organization, 41–45
    report table of contents, 43
    systematic deficiency in, 145
    top-down presentation, 44
    updates, 140
    worksheet example, 39
    worksheets, 38–52
Failure not verified (FNV), 72
Failure prevention, 1
    analytical approaches to, 37–61
    contributions to, 163
    cost of failure and, 167–82
    development phase and, 142
    in life cycle, 137–64
    practices, 120–29
    requirements, 120–22
    test, 122–26
    testing for, 63–85
    UML-based software development, 126–29
Failure probability, 1, 65, 66, 176
    aircraft electronics power supply, 211
    estimate uncertainties, 176
    nonorthogonal gyros, 217
    orthogonal gyros, 216
    reduction estimation, 175
    reduction in, 179
    for satellite subsystems, 196
Failure processing template, 146
Failure rate(s)
    base, 8
    channel processing, 214
    denominator, 16–18
    expression, 16–18
    flight control electronics, 213
    improved flight control electronics, 214
    numerator, 16
    published, 8
Failure Review Board (FRB), 72
Failure(s)
    analysis, 161–62
    aviation, 29–31

Failure(s) (continued)
  in calendar time, 17
  causes, 9
  Chernobyl, 27–29
  common threads, 32–34
  cost, 167–82
  developer's view, 16
  exposure to, 16
  histories, 143
  inevitability and, 21–22
  Mars spacecraft, 23–26
  nonchangeable, 147
  nonrandom, 96
  in operating time, 17
  organizational causes, 21–34
  probable cause summary, 33
  rare conditions for, 119
  rates, 88
  reporting, units of time, 18
  reporting system, 72
  software, 109–14
  space shuttle Columbia, 26–27
  telecommunications, 31–32
  thoroughly documented, 22–32
Failure/value (F/V) ratio, 180
Faults
  density, 112, 113
  exposure ratio, 113
  identification, 113
  interval for counting, 113
  removed, as found, 132
Fault tolerance, 129–32
  cost, 131
  dynamic, 131
  RBD for, 132
  static, 131
Fault tree analysis (FTA), 56–61
  application, 57
  basics, 57–58
  defined, 56–57
  example, 58–61
  as hierarchical procedure, 57–58
  missile detonation circuit, 58, 59
  symbols, 57
  See also Analytical approaches
Field failures
  causes, 8–12
  event classifications, 81
  investigations, 81–82

  not verified (FFNV), 145
  See also Failure(s)
Field test set (FTS), 82, 84
  time trends analysis, 84
  use of, 84
  See also In-service testing
First article acceptance test, 75
Fully automated implementation, 191, 192
Functional allocation, 150

General purpose computer (GPC), 102
Government Industry Data Exchange
       Program (GIDEP), 159, 162
Gyros, 215–17
  nonorthogonal orientation, 217–18
  orthogonal configurations, 215–17

In-house monitoring, 159–62
  awareness of technical information, 162
  failure reporting/analysis, 161–62
  quality assurance, 159–61
  See also Monitoring
In-service testing, 82–84
  built-in test, 83–84
  EDCs, 82–83
  FTS, 84
Interest groups, 3–4

k-out-of-n redundancy, 102–5
  capabilities, 104
  example, 103
  selection mechanism, 104
  symbols, 103

Life cycle
  activity blocks, 138
  activity elements, 137
  development phase, 137, 139
  failure cause introduction and, 163
  failure prevention in, 137–64
  format, 138–41
  milestones, 138
  phases, 137
  phases, reliability issues and, 141–48
  reliability and, 137
  spiral model, 139–40
  terminology, 138
  waterfall charts, 138, 139
Littlewood-Verall (L-V) model, 132

Maintenance
  budget, 193

effectiveness, increasing, 192–95
records, 193
repair actions, 193
Mars Climate Orbiter (MCO), 23
budget, 25
failure, 24
Mars Polar Lander (MPL), 23
budget, 25
failure cause, 23–24
sequence of operations, 24
See also Failure(s)
Mars spacecraft failures, 23–26
Mean-time-between-failures (MTBF), 7
assumptions on unavailability, 209
claim, 65
of dedicated power supply, 209–10
demonstrated, 64
ratio, 65
Mean-time-to-failure (MTTF), 7
Mean-time-to-repair (MTTR), 14
Medford switching center, 31
Mercury-Redstone launch, 52
Methods, 126–27
Milestones, 138
Model-based development, 49
Monitoring, 157–62
critical items, 157–62
in-house, 159–62
purchased items, 158–59
for reliability attainment, 159–62

Next highest level (NHL), 40, 41
N-modular redundancy (NMR), 99
Nonchangeable failures, 147
Nonoperating Reliability Databook, 98
Nonorthogonal gyros
failure probability, 217
orientation, 217–18
reference axes, 217
See also gyros
Normalized margins, 67

O&M phase, 137
FRACAS focus, 163
problem reports, 144–45
reliability issues in, 144–48
See also Life cycle
Operational capability (IOC), 75
Operational evaluation, 75–76
Optimum reliability, 167–70

concept, 167
starting point, 168
Organization, this book, 2–3
Organizational causes, 21–34
Orthogonal gyros
configurations, 215–16
failure probability, 216
reference axes, 216
See also Gyros

Pair-and-pair redundancy, 102
Pair-and-space redundancy, 101–2
benefits, 101
defined, 101
illustrated, 102
Parameter estimation, 8–9
Pareto distribution, 47, 77
Partial redundancy, 178
Partitioned redundancy, 90
Partitioning, 90–91
effect on failure probability, 91
Stratus, 91
Parts evaluation, 72–73
Perfective changes, 147
Plant simulation, 123–24
Poisson distribution, 6
Power outages
commercial, 185
duration plot, 186
fully automated alternative, 191, 192
notification function, 190
semiautomated alternative, 191–92
tolerating, 187
Power supplies, 200–210
alternative characteristics, 209
alternatives, 201–8
alternatives evaluation, 208–10
box replacement probability, 203–4
common, 201–6
dedicated, 206–8
location, 200
magnetic circuit breakers, 206
power diode, 205
protective components, 204, 205
selection framework, 200–201
See also Applications
Prism program, 9
Probability of failure, 1, 65, 66, 89
attitude determination, 218
calculation, 69

Probability of failure (continued)
 difference distribution and, 70
 due to test case, 126
 for low values of M, 68
 normalized margin vs., 67
 partitioning effect on, 91
 of TMR and simplex systems, 100
 See also Failure(s)
Problem reports, 144–45
 could not duplicate (CND), 145
 defined, 144
 field failure not verified (FFNV), 145
Process control deviations, 158–59
Process improvement, 174, 181
Production reliability acceptance tests (PRAT), 152

Quadruple redundancy, 106
Quadruple voting configurations, 102
Qualification test, 74–75
 data evaluation issues, 75
 defined, 74
 See also Tests
Quality assurance, 159–61
 control chart, 159–60
 statistical parameter presentation, 160
Quality of service (QoS), 183
Quantitative allocation, 150

Rapid prototyping, 141
Rare events (REs), 117
 causes, 117
 multiple, 117, 124, 126
 probability of, 125
 RR ratio, 124
Real-time software, 109
Receive HB method, 128
 counter method, 128
 failure mode, 128
 FMEA worksheet for, 129
 HB failure, 129
 use case diagram, 128
Redundancy, 175, 182
 added resource for, 178
 analytical, 105–6
 architecture comparison, 107
 at component level, 87–91
 dual, 91–98
 dynamic, 91–95
 higher-order configurations, 102–5
 k-out-of-n, 102–5

 pair-and-pair, 102
 pair-and-spare, 101–2
 partitioned, 90
 with power switching, 97
 quadruple, 106
 risk for, 175
 static, 91–95
 summary, 106
 techniques, 87–107
 temporal, 105
 triple, 98–102
 triple modular (TMR), 92, 99–101
Reliability
 aim, 1
 of aircraft electronics bay, 210–15
 assessment, 70
 change effects on, 147–48
 concept phase and, 141–42
 cost increment, 168
 data, 9
 defined, 6
 demonstration, 64–66
 development phase and, 137, 142–44
 engineering, 5–19
 of explosive device components, 8
 initial planning, 144
 life cycle and, 137
 nonmonetary resources and, 34
 of n-redundant component, 177
 O&M phase and, 144–48
 optimum, 167–70
 partial, 178
 qualification test (RQT), 152
 relevance of postdevelopment tests, 76–82
 relevance of tests during development, 70–76
 requirements, 151
 resources, 34
 software, 109–33
 testing, 64
 trends for redundant configurations, 88, 89
Reliability allocation, 150–51
 effort allocated to, 151
 functional, 150
 quantitative, 150
Reliability block diagrams (RBD), 9–12
 common power supplies, 203
 conventional, 12

defined, 9
    for fault tolerant software, 132
    for function display lighting, 10
    limitations, 12
    parallel structure, 10
    series relationship in, 10
Reliability development/growth test
        (RDGT), 152
Reliability improvement, 170
    expenditure, 172
    expenditure requirement, 170
    to meet QoS requirements, 183–92
    power supply analysis, 184–87
    server equipment availability, 187–92
Reliability program plan, 137, 148–52
    comment column, 148
    design and evaluation tasks, 150
    elements, 149
    requirement, 148
    tasks, 148
Repair(s)
    actions, 193
    electrical and pneumatic, 194
    rate formulations, 18–19
    time elements for, 19
    See also Maintenance
Reviews
    action items, 155
    decisions, 154–55
    defined, 152–53
    failure reports, 161
    first session, 154
    reliability, 153
    results, 154
    standards, 153
    topics, 155–56
Routine acceptance tests, 76–78
    concern identification, 78
    defined, 76
    See also Tests
RPN, 51–52
    computation, 52
    difficulty of detection, 51
    high, 51
    highest possible, 52

Safety critical loads, 211–13
    partitioned, 212
    single channel data, 212
    See also Aircraft electronics bay

Safety margin approach, 68
Salt Lake City central office, 31–32
Screening, 175, 182
Screening tests, 78–81
    cordless phone example, 79
    defined, 78
    frequency distribution and,, 79, 80
    infant mortality example, 79–81
    use of, 80–81
Semiautomatic implementation, 191–92
Service equipment
    availability, 187–92
    block diagram for availability, 189
    installation, 188
    modified installation, 191
Severity categories, 41
Sneak Circuit Analysis Handbook, 52
Sneak circuit analysis (SCA), 52–56
    basics, 52–55
    cargo door latch function, 53
    circuit patterns, 54
    defined, 52
    observations, 53
    positive/negative power sources and, 55
    techniques, 55–56
    See also Analytical approaches
Software
    change requests summary, 110, 111
    embedded, 109
    failure prevention practices, 120–29
    failures, 109–14
    monitoring, 114
    real-time, 109
Software fault tolerance, 129–32
    cost, 131
    dynamic, 131
    RBD for, 132
    static, 131
Software reliability, 109–33
    models, 132
    summary, 132–33
Software testing, 114–20
    limitations, 115
    results, 115
Spacecraft attitude determination, 215–18
    difficulties, 215
    nonorthogonal gyro orientation, 217–18
    orthogonal gyro configurations, 215–17
    See also Applications

Space Shuttle Avionics (SSA) software, 116
Space shuttle Columbia, 26–27
   investigation, 26
   organizational causes, 26–27
   See also Failure(s)
Spiral model, 139–40
   defined, 139
   for development phase, 140
   See also Life cycle
State transition diagrams
   defined, 13
   illustrated, 14
   uses, 14
State transition methods, 12–16
Static redundancy, 91–95
   defined, 91
   illustrated, 92
   for mechanical elements, 92
   software application, 129, 131
   See also Redundancy
System failure effects summary, 44
System integration test, 74
System reliability. *See* Reliability

Telecommunications, 31–32
   Medford switching facility, 31
   Salt Lake City central office, 31–32
Temporal redundancy, 105
Testing, 63–85, 152
   by attributes, 70
   code-based, 122
   for contractual purposes, 64
   cost, 63, 152
   for design margins, 63
   early, 63
   in-service, 82–84
   random, 122
   reliability, 64
   requirements-based, 122
   software, 114–20
   summary, 85
   by variables, 70
Tests
   breadboard, 73
   component prototype, 74
   during development, 70–76
   environmental stress screening (ESS), 152
   first article acceptance, 75
   plan, 71

postdevelopment, 76–82
procedure, 71
production reliability acceptance (PRAT), 152
purpose, 73
qualification, 74–75
reliability development/growth (RDGT), 152
reliability qualification (RQT), 152
reliability relevance of, 70–82
report, 71
results, quantitative recording of, 73
routine acceptance, 76–78
screening, 78–81
selection, 63
specification, 71
system integration, 74
typical, 73
Time to restore service (TTRS), 18
   in availability calculations, 18
   defined, 18
Triple modular redundancy (TMR), 92, 99–101
   applications, 100
   defined, 99
   failure probability, 100
   with voting, 99
   See also Redundancy
Triple redundancy, 98–102
   pair-and-space, 101–2
   TMR, 99–101
   See also Redundancy

UML-based software development, 126–29
   tools, 126
   use case diagram, 127–28
Uniform Modeling Language (UML), 49
Uninterruptible power supply (UPS), 184
Use case diagrams, 127–28
   active/standby, 127
   defined, 127
   receive HB, 128

Waterfall charts, 138–39
   defined, 138
   for development phase, 139
   See also Life cycle
Work breakdown structure (WBS), 148

Recent Titles in the Artech House Technology Management and Professional Development Library

Bruce Elbert, Series Editor

Advanced Systems Thinking, Engineering, and Management, Derek K. Hitchins

Critical Chain Project Management, Lawrence P. Leach

Decision Making for Technology Executives: Using Multiple Perspectives to Improve Performance, Harold A. Linstone

Designing the Networked Enterprise, Igor Hawryszkiewycz

Engineering and Technology Management Tools and Applications, B. S. Dhillon

The Entrepreneurial Engineer: Starting Your Own High-Tech Company, R. Wayne Fields

Evaluation of R&D Processes: Effectiveness Through Measurements, Lynn W. Ellis

From Engineer to Manager: Mastering the Transition, B. Michael Aucoin

Introduction to Information-Based High-Tech Services, Eric Viardot

Introduction to Innovation and Technology Transfer, Ian Cooke and Paul Mayes

ISO 9001:2000 Quality Management System Design, Jay Schlickman

Managing Complex Technical Projects: A Systems Engineering Approach, R. Ian Faulconbridge and Michael J. Ryan

Managing Engineers and Technical Employees: How to Attract, Motivate, and Retain Excellent People, Douglas M. Soat

Managing Successful High-Tech Product Introduction, Brian P. Senese

Managing Virtual Teams: Practical Techniques for High-Technology Project Managers, Martha Haywood

Mastering Technical Sales: The Sales Engineer's Handbook, John Care and Aron Bohlig

The New High-Tech Manager: Six Rules for Success in Changing Times, Kenneth Durham and Bruce Kennedy

Planning and Design for High-Tech Web-Based Training, David E. Stone and Constance L. Koskinen

Preparing and Delivering Effective Technical Presentations, Second Edition, David Adamy

Reengineering Yourself and Your Company: From Engineer to Manager to Leader, Howard Eisner

The Requirements Engineering Handbook, Ralph R. Young

Running the Successful Hi-Tech Project Office, Eduardo Miranda

Successful Marketing Strategy for High-Tech Firms, Second Edition, Eric Viardot

Successful Proposal Strategies for Small Businesses: Using Knowledge Management to Win Government, Private Sector, and International Contracts, Third Edition, Robert S. Frey

Systems Approach to Engineering Design, Peter H. Sydenham

Systems Engineering Principles and Practice, H. Robert Westerman

Systems Reliability and Failure Prevention, Herbert Hecht

Team Development for High-Tech Project Managers, James Williams

For further information on these and other Artech House titles, including previously considered out-of-print books now available through our In-Print-Forever® (IPF®) program, contact:

Artech House
685 Canton Street
Norwood, MA 02062
Phone: 781-769-9750
Fax: 781-769-6334
e-mail: artech@artechhouse.com

Artech House
46 Gillingham Street
London SW1V 1AH UK
Phone: +44 (0)20 7596-8750
Fax: +44 (0)20 7630-0166
e-mail: artech-uk@artechhouse.com

Find us on the World Wide Web at:
www.artechhouse.com